# FANTASIA MATHEMATICA

# FANTASIA MATHEMATICA

Being a set of stories,
together with a group of oddments and diversions,
all drawn from the universe of mathematics.

Compiled and edited by

Clifton Fadiman

COPERNICUS
AN IMPRINT OF SPRINGER-VERLAG

Published in 1997 by Copernicus, an imprint of
Springer-Verlag New York, Inc.

This edition is reprinted by arrangement with Simon & Schuster, Inc.



All characters except for historical personages are fictitious.

Copernicus Books
37 East 7th Street
New York, NY 10003
www.copernicusbooks.com

Library of Congress Cataloging-in-Publication Data

Fantasia mathematica : being a set of stories, together with a group of oddments and diversions, all drawn from the universe of mathematics / assembled and edited, with an introduction by Clifton Fadiman
p. cm.
Originally published: New York : Simon and Schuster, 1958.
ISBN 0-387-94931-3 (softcover : alk. paper)
1. Mathematics—Literary collections. I. Fadiman, Clifton, 1904–.
PN6071.M3F35 1997 97-7654
808.8'0356—dc21

Cover: Ilya Bolotowsky, *Surreal Dream*, 1933, oil on canvas, 16" x 24".


Manufactured in the United States of America.
Printed on acid-free paper.

9 8 7 6 5

ISBN 0-387-94931-3

FOR

*Scott Buchanan,*

MOVER OF MINDS,

WHO STARTED THIS

# *CONTENTS*

## II. IMAGINARIES

Contents

# *INTRODUCTION*

ABOUT twenty-five years ago I spent a weekend at the country home of the poet and critic Mark Van Doren. Among my fellow-guests was an old friend of my college days, Scott Buchanan, to whom this collection is dedicated and who was then and remains today one of the most remarkable men I have ever met. I should like some day to write at length about him. He is the only distinguished representative I know of a species practically extinct, or in any case obsolescent—the vagabond intellectual. I do not mean that he is a bohemian, only that his life has been guided by the behests of his curious mind rather than strait-jacketed by the usual limitations of a "career." He has had a brilliant career nevertheless, but only as a by-product. Though he is a philosopher by training, his real interest is simply in thought, thought in almost any area. Wherever intellectual curiosity has led him, there he has strayed, as reveler or as ponderer. He has been a professor, a writer, a dabbler in politics, a sociological analyst, many other things. But at bottom he is an itinerant mind, not unlike those attractive wandering students of the twelfth century who have immortalized themselves in the Goliardic Songs.

Among other things he is a good mathematician. Should you ever come across his out-of-print Poetry and Mathematics you will find its witty and occasionally obscure pages more than rewarding. On this particular sunny afternoon he amused us by the display of a few topological tricks which have long been familiar both to mathematicians and to stage magicians, but which were new to us. One of these tricks involved the Moebius strip, exploited by several stories in this volume, and clearly described and explained by Mr. Gardner in his delightful "No-Sided Professor."

It was at this point that I suddenly realized that I was interested in mathematics. An indifferent student of the subject in high school, I had not thought about mathematics for years, feeling it to be beyond my mental capacities. But Mr. Buchanan's parlor trick, with its revelation of the fact, for example, that a surface can exist with only one side and one edge (try it yourself) aroused a curiosity which I have been gratifying ever since. I still know no real mathematics; but I know what mathematics is about, and something of the way in which the mathematical mind seems to work, and a fair amount about the lives of the great mathematicians. I have written elsewhere, in an essay called "Meditations of a Mathematical Moron,"[1] of the pleasures of reading about mathematics, pleasures open to anyone who has received a conventional secondary-school education, and is willing to do a little mental work.

One of the by-products of this interest is now in your hands. For about five years I have been collecting stories and imaginative oddments, often trivial, about mathematics. The stories are suitable not for mathematicians, who will be bored by their naïveté, but for non-mathematicians capable of being amused or surprised by the devious connections between the imagination and what would appear to be a discipline of the utmost rigor. You will not learn much mathematics from them—they are intended to amuse or tease rather than to instruct; but they may lead readers like

[1] See Any Number Can Play, World Publishing Company, New York, 1957.

*myself, curious but unlearned, to create a better image of a few mathematical ideas.*

*There is not a great deal of imaginative literature, at any rate in the languages with which I am acquainted, involving mathematics. This may be because literary men assume—wrongly—that the subject is either too difficult or too refractory for their purposes. And it is true that mathematics is not "visual" in the sense that astronomy and to some extent biology and chemistry are; so that the mythopoeic mind would seem to have little to work on. Nevertheless this volume manages to salvage what I dare think to be some interesting samples of what mathematics can yield to the poet, the humorist and the fiction writer. As I am positive, however, that in my ignorance I have overlooked many other examples, I would be glad to hear about them from more learned readers.*

*The anthology is highly scalene, for its three parts are of unequal length. The first section consists of seven stories, dialogues, or excerpts from novels. They are by good writers—in the case of Plato, by one of the greatest. But as (with the exception of Plato) they are not by writers ordinarily associated with the subject, I have called them Odd Numbers.*

*The first story, a most moving and beautiful one, has been frequently anthologized, and I must assume its familiarity to most readers. Yet this collection could not well dispense with "Young Archimedes." It is the best modern attempt I know to convey to the reader the peculiar preciousness and value to mankind of mathematical genius. It is not written by a mathematician, but by a writer trained, perhaps overtrained, in the humanities; and so to a degree it is written from the outside. We are stricken by the tragedy of the little hero, but our sympathy does not flow from a comprehension of the actual content of his mind, only from the waste and futility of his brief life. For all that, a wonderful story.*

*Most of the other Odd Numbers require little comment. The*

contribution from Mr. Cabell is, of course, an elaborate naughty joke, and a delightful one, though it holds up an ideal of mathematical performance that few of us can hope to achieve in actual life. Still, a man's reach . . .

I have included the larger part of Plato's "Meno," partly because it expresses so concretely his wild doctrine of reminiscence and partly because the mathematics in it makes sense—his other references to the discipline are ordinarily saturated with a, to me, unpalatable mysticism.

The second section of this anthology, entitled Imaginaries, is the largest, comprising sixteen science-fiction stories plus a commentary on one of them by the learned rocket and space-travel expert, Willy Ley. All of these stories actually have a certain amount of mathematical thinking in them; or else they make fun of mathematical thinking. But they are not para-mathematical stories—that is, they are not concerned with the fancies of time travel or parallel worlds, or similar science-fiction stand-bys. Or, when they are, as with "—And He Built a Crooked House," the mathematics is by no means completely unsophisticated.

I make no claim to excessive originality; science-fiction buffs will recognize several old favorites, including such masterpieces in their restricted field as the tales by Heinlein, Clarke, and Gardner, who is probably the best actual mathematician of the lot, and from whose amusing topological fables much can be learned. A few items may be unfamiliar even to the professionals. Edward Page Mitchell's "The Tachypomp" is a curio. It goes back to 1873, at which date H. G. Wells was only seven years old. In its now quaintly old-fashioned way it foreshadows many of the themes and situations that were to be exploited by science-fictioneers more than a generation later. Kurd Lasswitz's "The Universal Library" appears here at the suggestion of Willy Ley, its translator. It has not heretofore been published in this country, though it is famous in its homeland, Germany. It is a deadpan answer to the question, How long would it take those monkeys, pounding away at their typewriters, to come up by chance with the works of Shakespeare? Mr. Ley's commentary on this remarkable story is, as one would expect, both interesting and scholarly.

Several of the stories are minor classics. For example, I understand that Arthur C. Clarke's "Superiority" is required reading in certain classes at M.I.T.; and the two tales of Martin Gardner are surely superlative translations on to the plane of entertainment of certain basic topological concepts.

The reader will note that several of the stories base themselves upon our old friend, the Moebius strip. It is remarkable how many narrative variations are possible using only this particular device; I have in my files half a dozen other Moebius strip stories, all interesting, though perhaps not quite so amusing as the item by Mr. Upson or so troubling as the one by Mr. Deutsch.

Topology, however, is only one of the mathematical fields touched upon in the Imaginaries section. The reader will find himself making some superficial acquaintance with Fermat's Last Theorem, with the fourth dimension, communication theory, various concepts of infinity, the theory of numbers, and the fantastic principles governing compound interest. I do not claim that these stories hold any great profundity, but I do claim that they are first-rate recreational mathematics, and in a few cases highly ingenious specimens of narrative art.

The final section of Fantasia Mathematica, called Fractions, is simply an unsystematic assemblage of rhymes, jingles, serious short poems, fancies, jokes, fables and anecdotes that touch on mathematics. Many of them are pleasant trivialities and are intended to amuse, nothing more. Yet it is noteworthy that their authors should include names as illustrious as Eddington, Lewis Carroll, Marvell, Dekker, Arthur Schnitzler, Edna Millay and Carl Sandburg. One of these days I should like to make a whole *book of oddities connected with mathematics and its fascinating history. If I ever do I shall make sure to include the fact that in* 1899 *the Indiana Legislature very nearly passed a bill providing that in all Indiana schools pi should be* 4 *instead of* 3.14159 . . .

*Though the idea of this book was conceived and much of its material gathered some years before the publication of James Newman's magnificent four-volume anthology* The World of Mathematics, *I am frank in stating that I was in part encouraged to publish it by the extraordinary reception given Mr. Newman's work. Many years ago, when I was a publisher's editor, I felt vaguely that there existed in our country a non-professional audience for mathematics. I still feel a certain pride in having helped to midwife E. T. Bell's now classic* Men of Mathematics, *probably the best biographical writing on the subject to be found in English. That book, published as long ago as* 1937, *was successful, and is still in demand. But it is apparent, from Mr. Newman's triumph, that during the last twenty years an entirely new audience for popular works touching on mathematics has developed. I do not think it fantastic to say that science-fiction is in part responsible for this renewal of interest in an age-old discipline. The mind can be stimulated by fancies as well as by rigorous thought, and I hope that many a high-school student who is legitimately bored to death by the mathematics taught in many of our classrooms will find himself seduced into genuine mathematical curiosity by the thought-provoking stories of Huxley and Capek, the whimsy of Russell Maloney, the humor of Martin Gardner, the troubling problems posed by Arthur C. Clarke.*

*There is a certain guilt all veteran anthologizers (and I fear I am one of them) feel. Anthological guilt, we might call it. All anthologies are thefts, of course, and only a few, such as Mr. Newman's, are truly creative. In many cases, they are thefts of thefts.* Fantasia Mathematica *does not fill me with quite as much an-*

thological guilt as other collections I have made. Its subject is fresh, and I think the reader, unless he is already a good mathematician, will find his imagination stimulated and his mind set going in new directions. At any rate, I have found its compilation, which began as a labor of love, a great deal of fun, and I hope the readers of Fantasia Mathematica will derive from it some sense of the wonder, the oddity and even the humor of the most magnificent, as well as the most basic, of all the strange sciences man has invented, discovered, or—if you are to believe Plato—remembered.

For help and suggestions I am grateful to the following: Groff Conklin; August Derleth; Howard DeVore; Martin Gardner; Willy Ley; H. Nearing, Jr.; Professor William L. Schaaf; William Sloane; George Stevens; Anne Whitmore; and Alan Williams.

—Clifton Fadiman
August 1957

"Dost thou never study the mathematics?"
—WEBSTER, *The Duchess of Malfi*

# I

# ODD NUMBERS

ALDOUS HUXLEY

# Young Archimedes

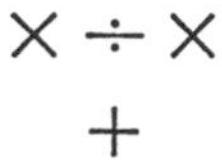

IT WAS THE VIEW which finally made us take the place. True, the house had its disadvantages. It was a long way out of town and had no telephone. The rent was unduly high, the drainage system poor. On windy nights, when the ill-fitting panes were rattling so furiously in the window frames that you could fancy yourself in an hotel omnibus, the electric light, for some mysterious reason, used invariably to go out and leave you in the noisy dark. There was a splendid bathroom; but the electric pump, which was supposed to send up water from the rainwater tanks in the terrace, did not work. Punctually every autumn the drinking well ran dry. And our landlady was a liar and a cheat.

But these are the little disadvantages of every hired house, all over the world. For Italy they were not really at all serious. I have seen plenty of houses which had them all and a hundred others, without possessing the compensating advantages of ours—the southward-facing garden and terrace for the winter and spring, the large cool rooms against the midsummer heat, the hilltop air and freedom from mosquitoes, and finally the view.

And what a view it was! Or rather, what a succession of views. For it was different every day; and without stirring from the house one had the impression of an incessant change of scene: all the delights of travel without its fatigues. There were autumn days when all the valleys were filled with mist and the crest of the Apennines rose darkly out of a flat white lake. There were days when the mist invaded even our hilltop and we were enveloped in a soft vapor in which the mist-colored olive trees that sloped away below our windows toward the valley disappeared as though into their own spiritual essence; and the only firm and definite things in the small, dim world within which we found ourselves confined were the two tall black cypresses growing on a little projecting terrace a hundred feet down the hill. Black, sharp, and solid, they stood there, twin pillars of Hercules at the extremity of the known universe; and beyond them there was only pale cloud and round them only the cloudy olive trees.

These were the wintry days; but there were days of spring and autumn, days unchallengingly cloudless, or—more lovely still—made various by the huge floating shapes of vapor that, snowy above the faraway, snow-capped mountains, gradually unfolded, against the pale bright blue, enormous heroic gestures. And in the height of the sky the bellying draperies, the swans, the aerial marbles, hewed and left unfinished by gods grown tired of creation almost before they had begun, drifted sleeping along the wind, changing form as they moved. And the sun would come and go behind them; and now the town in the valley would fade and almost vanish in the shadow, and now, like an immense fretted jewel between the hills, it would glow as though by its own light. And looking across the nearer tributary valley that wound from below our crest down toward the Arno, looking over the low dark shoulder of hill on whose extreme promontory stood the towered church of San Miniato, one saw the huge dome airily hanging on its ribs of masonry, the square campanile, the sharp spire of Santa Croce, and the canopied tower of the Signoria, rising above the intricate maze of houses, distinct and brilliant, like small treasures carved out of precious stones. For a moment only, and then their light would fade away once more, and the traveling beam would pick out, among the indigo hills beyond, a single golden crest.

There were days when the air was wet with passed or with approaching rain, and all the distances seemed miraculously near and clear. The olive trees detached themselves one from another on the distant slopes; the faraway villages were lovely and pathetic, like the most exquisite small toys. There were days in summertime, days of impending thunder when, bright and sunlit against huge bellying masses of black and purple, the hills and the white houses shone as it were precariously, in a dying splendor, on the brink of some fearful calamity.

How the hills changed and varied! Every day and every hour of the day, almost, they were different. There would be moments when, looking across the plains of Florence, one would see only a dark blue silhouette against the sky. The scene had no depth; there was only a hanging curtain painted flatly with the symbols of the mountains. And then, suddenly almost, with the passing of a cloud, or when the sun had declined to a certain level in the sky, the flat scene transformed itself; and where there had been only a painted curtain, now there were ranges behind ranges of hills, graduated tone after tone from brown, or gray, or a green gold to faraway blue. Shapes that a moment before had been fused together indiscriminately into a single mass now came apart into their constituents. Fiesole, which had seemed only a spur of Monte Morello, now revealed itself as the jutting headland of another system of hills, divided from the nearest bastions of its greater neighbor by a deep and shadowy valley.

At noon, during the heats of summer, the landscape became dim, powdery, vague and almost colorless under the midday sun; the hills disappeared into the trembling fringes of the sky. But as the afternoon wore on the landscape emerged again, it dropped its anonymity, it climbed back out of nothingness into form and life. And its life, as the sun sank and slowly sank through the long afternoon, grew richer, grew more intense with every moment. The level light, with its attendant long, dark shadows, laid bare, so to speak, the anatomy of the land; the hills—each western escarpment shining, and each slope averted from the sunlight profoundly shadowed—became massive, jutty, and solid. Little folds of dimples in the seemingly even ground revealed themselves. Eastward from our hilltop, across the plain of the Ema, a great bluff cast its ever-increasing shadow; in the surrounding brightness of the valley a whole town lay eclipsed within it. And as the sun expired on the horizon, the further hills flushed in its warm light, till their illumined flanks were the color of tawny roses; but the valleys were already filled with the blue mist of the evening. And it mounted, mounted; the fire went out of the western windows of the populous slopes; only the crests were still alight, and at last they too were all extinct. The mountains faded and fused together again into a flat painting of mountains against the pale evening sky. In a little while it was night; and if the moon were full, a ghost of the dead scene still haunted the horizons.

Changed in its beauty, this wide landscape always preserved a quality of humanness and domestication which made it, to my mind at any rate, the best of all landscapes to live with. Day by day one traveled through its different beauties; but the journey, like our ancestors' Grand Tour, was always a journey through civilization. For all its mountains, its deep slopes and deep valleys, the Tuscan scene is dominated by its inhabitants. They have cultivated every rood of ground that can be cultivated; their houses are thickly scattered even over the hills, and the valleys are populous. Solitary on the hilltop, one is not alone in a wilderness. Man's traces are across the country, and already—one feels it with satisfaction as one looks out across it—for centuries, for thousands of years, it has been his, submissive, tamed, and humanized. The wide, blank moorlands, the sands, the forests of innumerable trees—these are places for occasional visitation, healthful to the spirit which submits itself to them for not too long. But fiendish influences as well as divine haunt these total solitudes. The vegetative life of plants and things is alien and hostile to the human. Men cannot live at ease except where they have mastered their surroundings and where their accumulated lives outnumber and outweigh the vegetative lives about them. Stripped of its dark wood, planted, terraced and tilled almost to the mountains' tops, the Tuscan landscape is humanized and safe. Sometimes upon those who live in the midst of it there comes a longing for some place that is solitary, inhuman, lifeless, or peopled only with alien life. But the longing is soon satisfied, and one is glad to return to the civilized and submissive scene.

I found that house on the hilltop the ideal dwelling place. For there, safe in the midst of a humanized landscape, one was yet alone; one could be as solitary as one liked. Neighbors whom one never sees at close quarters are the ideal and perfect neighbors.

Our nearest neighbors, in terms of physical proximity, lived very near. We had two sets of them, as a matter of fact, almost in the same house with us. One was the peasant family, who lived in a long, low building, part dwelling house, part stables, storerooms and cow sheds, adjoining the villa. Our other neighbors—intermittent neighbors, however, for they only ventured out of town every now and then, during the most flawless weather—were the owners of the villa, who had reserved for themselves the smaller wing of the huge L-shaped house—a mere dozen rooms or so—leaving the remaining eighteen or twenty to us.

They were a curious couple, our proprietors. An old husband, gray, listless, tottering, seventy at least; and a signora of about forty, short, very plump, with tiny fat hands and feet and a pair of very large, very dark black eyes, which she used with all the skill of a born comedian. Her vitality, if you could have harnessed it and made it do some useful work, would have supplied a whole town with electric light. The physicists talk of deriving energy from the atom; they would be more profitably employed nearer home—in discovering some way of tapping those enormous stores of vital energy which accumulate in unemployed women of sanguine temperament and which, in the present imperfect state of social and scientific organization, vent themselves in ways that are generally so deplorable in interfering with other people's affairs, in working up emotional scenes, in thinking about love and making it, and in bothering men till they cannot get on with their work.

Signora Bondi got rid of her superfluous energy, among other ways, by "doing in" her tenants. The old gentleman, who was a retired merchant with a reputation for the most perfect rectitude, was allowed to have no dealings with us. When we came to see the house, it was the wife who showed us round. It was she who, with a lavish display of charm, with irresistible rollings of the eyes, expatiated on the merits of the place, sang the praises of the electric pump, glorified the bathroom (considering which, she insisted, the rent was remarkably moderate), and when we suggested calling in a surveyor to look over the house, earnestly begged us, as though our well-being were her only consideration, not to waste our money unnecessarily in doing anything so superfluous. "After all," she said, "we are honest people. I wouldn't dream of letting you the house except in perfect condition. Have confidence." And she looked at me with an appealing, pained expression in her magnificent eyes, as though begging me not to insult her by my coarse suspiciousness. And leaving us no time to pursue the subject of surveyors any further, she began assuring us that our little boy was the most beautiful angel she had ever seen. By the time our interview with Signora Bondi was at an end, we had definitely decided to take the house.

"Charming woman," I said, as we left the house. But I think that Elizabeth was not quite so certain of it as I.

Then the pump episode began.

On the evening of our arrival in the house we switched on the electricity. The pump made a very professional whirring noise; but no water came out of the taps in the bathroom. We looked at one another doubtfully.

"Charming woman?" Elizabeth raised her eyebrows.

We asked for interviews; but somehow the old gentleman could never see us, and the Signora was invariably out or indisposed. We left notes; they were never answered. In the end, we found that the only method of communicating with our landlords, who were living in the same house with us, was to go down into Florence and send a registered express letter to them. For this they had to sign two separate receipts and even, if we chose to pay forty centimes more, a third incriminating document, which was then returned to us. There could be no pretending, as there always was with ordinary letters or notes, that the communication had never been received. We began at last to get answers to our complaints. The Signora, who wrote all the letters, started by telling us that, naturally, the pump didn't work, as the cisterns were empty, owing to the long drought. I had to walk three miles to the post office in order to register my letter reminding her that there had been a violent thunderstorm only last Wednesday, and that the tanks were consequently more than half full. The answer came back: bath water had not been guaranteed in the contract; and if I wanted it, why hadn't I had the pump looked at before I took the house? Another walk into town to ask the Signora next door whether she remembered her adjurations to us to have confidence in her, and to inform her that the existence in a house of a bathroom was in itself an implicit guarantee of bath water. The reply to that was that the Signora couldn't continue to have communications with people who wrote so rudely to her. After that I put the matter into the hands of a lawyer. Two months later the pump was actually replaced. But we had to serve a writ on the lady before she gave in. And the costs were considerable.

One day, toward the end of the episode, I met the old gentleman in the road, taking his big Maremman dog for a walk—or being taken, rather, for a walk by the dog. For where the dog pulled the old gentleman had perforce to follow. And when it stopped to smell, or scratch the ground, or leave against a gatepost its visiting card or an offensive challenge, patiently, at his end of the leash, the old man had to wait. I passed him standing at the side of the road, a few hundred yards below our house. The dog was sniffing at the roots of one of the twin cypresses which grew one on either side of the entry to a farm; I heard the beast growling indignantly to itself, as though it scented an intolerable insult. Old Signor Bondi, leashed to his dog, was waiting. The knees inside the tubular gray trousers were slightly bent. Leaning on his cane, he stood gazing mournfully and vacantly at the view. The whites of his old eyes were discolored, like ancient billiard balls. In the gray, deeply wrinkled face, his nose was dyspeptically red. His white mustache, ragged and yellowing at the fringes, drooped in a melancholy curve. In his black tie he wore a very large diamond; perhaps that was what Signora Bondi had found so attractive about him.

I took off my hat as I approached. The old man stared at me absently, and it was only when I was already almost past him that he recollected who I was.

"Wait," he called after me, "wait!" And he hastened down the road in pursuit. Taken utterly by surprise and at a disadvantage—for it was engaged in retorting to the affront imprinted on the cypress roots—the dog permitted itself to be jerked after him. Too much astonished to be anything but obedient, it followed its master. "Wait!"

I waited.

"My dear sir," said the old gentleman, catching me by the lapel of my coat and blowing most disagreeably in my face, "I want to apologize." He looked around him, as though afraid that even here he might be overheard. "I want to apologize," he went on, "about the wretched pump business. I assure you that, if it had been only my affair, I'd have put the thing right as soon as you asked. You were quite right: a bathroom is an implicit guarantee of bath water. I saw from the first that we should have no chance if it came to court. And besides, I think one ought to treat one's tenants as handsomely as one can afford to. But my wife—" he lowered his voice—"the fact is that she likes this sort of thing, even when she knows that she's in the wrong and must lose. And besides, she hoped, I dare say, that you'd get tired of asking and have the job done yourself. I told her from the first that we ought to give in; but she wouldn't listen. You see, she enjoys it. Still, now she sees that it must be done. In the course of the next two or three days you'll be having your bath water. But I thought I'd just like to tell you how . . ." But the Maremmano, which had recovered by this time from its surprise of a moment since, suddenly bounded, growling, up the road. The old gentleman tried to hold the beast, strained at the leash, tottered unsteadily, then gave way and allowed himself to be dragged off. ". . . how sorry I am," he went on, as he receded from me, "that this little misunderstanding . . ." But it was no use. "Good-by." He smiled politely, made a little deprecating gesture, as though he had suddenly remembered a pressing engagement, and had no time to explain what it was. "Good-by." He took off his hat and abandoned himself completely to the dog.

A week later the water really did begin to flow, and the day after our first bath Signora Bondi, dressed in dove-gray satin and wearing all her pearls, came to call.

"Is it peace now?" she asked, with a charming frankness, as she shook hands.

We assured her that, so far as we were concerned, it certainly was.

"But why did you write me such dreadfully rude letters?" she said, turning on me a reproachful glance that ought to have moved the most ruthless malefactor to contrition. "And then that writ. How could you? To a lady . . ."

I mumbled something about the pump and our wanting baths.

"But how could you expect me to listen to you while you were in that mood? Why didn't you set about it differently—politely, charmingly?" She smiled at me and dropped her fluttering eyelids.

I thought it best to change the conversation. It is disagreeable, when one is in the right, to be made to appear in the wrong.

A few weeks later we had a letter—duly registered and by express messenger—in which the Signora asked us whether we proposed to renew our lease (which was only for six months), and notifying us that, if we did, the rent would be raised twenty-five per cent, in consideration of the improvements which had been carried out. We thought ourselves lucky, at the end of much bargaining, to get the lease renewed for a whole year with an increase in the rent of only fifteen per cent.

It was chiefly for the sake of the view that we put up with these intolerable extortions. But we had found other reasons, after a few days' residence, for liking the house. Of these the most cogent was that, in the peasant's youngest child, we had discovered what seemed the perfect playfellow for our own small boy. Between little Guido—for that was his name—and the youngest of his brothers and sisters there was a gap of six or seven years. His two older brothers worked with their father in the fields; since the time of the mother's death, two or three years before we knew them, the eldest sister had ruled the house, and the younger, who had just left school, helped her and in between-whiles kept an eye on Guido, who by this time, however, needed very little looking after; for he was between six and seven years old and as precocious, self-assured, and responsible as the children of the poor, left as they are to themselves almost from the time they can walk, generally are.

Though fully two and a half years older than little Robin—and at that age thirty months are crammed with half a lifetime's experience —Guido took no undue advantage of his superior intelligence and strength. I have never seen a child more patient, tolerant, and untyrannical. He never laughed at Robin for his clumsy efforts to imitate his own prodigious feats; he did not tease or bully, but helped his small companion when he was in difficulties and explained when he could not understand. In return, Robin adored him, regarded him as a model and perfect Big Boy, and slavishly imitated him in every way he could.

These attempts of Robin's to imitate his companion were often exceedingly ludicrous. For by an obscure psychological law, words and actions in themselves quite serious become comic as soon as they are copied; and the more accurately, if the imitation is a deliberate parody, the funnier—for an overloaded imitation of someone we know does not make us laugh so much as one that is almost indistinguishably like the original. The bad imitation is only ludicrous when it is a piece of sincere and earnest flattery which does not quite come off. Robin's imitations were mostly of this kind. His heroic and unsuccessful attempts to perform the feats of strength and skill, which Guido could do with ease, were exquisitely comic. And his careful, long-drawn imitations of Guido's habits and mannerisms were no less amusing. Most ludicrous of all, because most earnestly undertaken and most incongruous in the imitator, were Robin's impersonations of Guido in a pensive mood. Guido was a thoughtful child given to brooding and sudden abstractions. One would find him sitting in a corner by himself, chin in hand, elbow on knee, plunged, to all appearances, in the profoundest meditation. And sometimes, even in the midst of his play, he would suddenly break off, to stand, his hands behind his back, frowning and staring at the ground. When this happened, Robin became overawed and a little disquieted. In a puzzled silence he looked at his companion. "Guido," he would say softly, "Guido." But Guido was generally too much preoccupied to answer; and Robin, not venturing to insist, would creep near him, and throwing himself as nearly as possible into Guido's attitude—standing Napoleonically, his hands clasped behind him, or sitting in the posture of Michelangelo's Lorenzo the Magnificent—would try to meditate too. Every few seconds he would turn his bright blue eyes toward the elder child to see whether he was doing it quite right. But at the end of a minute he began to grow impatient; meditation wasn't his strong point. "Guido," he called again and, louder, "Guido!" And he would take him by the hand and try to pull him away. Sometimes Guido roused himself from his reverie and went back to the interrupted game. Sometimes he paid no attention. Melancholy, perplexed, Robin had to take himself off to play by himself. And Guido would go on sitting or standing there, quite still; and his eyes, if one looked into them, were beautiful in their grave and pensive calm.

They were large eyes, set far apart and, what was strange in a dark-haired Italian child, of a luminous pale blue-gray color. They were not always grave and calm, as in these pensive moments. When he was playing, when he talked or laughed, they lit up; and the surface of those clear, pale lakes of thought seemed, as it were, to be shaken into brilliant sun-flashing ripples. Above those eyes was a beautiful forehead, high and steep and domed in a curve that was like a subtle curve of a rose petal. The nose was straight, the chin small and rather pointed, the mouth drooped a little sadly at the corners.

I have a snapshot of the two children sitting together on the parapet of the terrace. Guido sits almost facing the camera, but looking a little to one side and downward; his hands are crossed in his lap and his expression, his attitude are thoughtful, grave, and meditative. It is Guido in one of those moods of abstraction into which he would pass even at the height of laughter and play—quite suddenly and completely, as though he had all at once taken it into his head to go away and left the silent and beautiful body behind, like an empty house, to wait for his return. And by his side sits little Robin, turning to look up at him, his face half averted from the camera, but the curve of his cheek showing that he is laughing; one little raised hand is caught at the top of a gesture, the other clutches at Guido's sleeve, as though he were urging him to come away and play. And the legs dangling from the parapet have been seen by the blinking instrument in the midst of an impatient wriggle; he is on the point of slipping down and running off to play hide-and-seek in the garden. All the essential characteristics of both the children are in that little snapshot.

"If Robin were not Robin," Elizabeth used to say, "I could almost wish he were Guido."

And even at that time, when I took no particular interest in the child, I agreed with her. Guido seemed to me one of the most charming little boys I had ever seen.

We were not alone in admiring him. Signora Bondi when, in those cordial intervals between our quarrels, she came to call, was constantly speaking of him. "Such a beautiful, beautiful child!" she would exclaim with enthusiasm. "It's really a waste that he should belong to peasants who can't afford to dress him properly. If he were mine, I should put him into black velvet; or little white knickers and a white knitted silk jersey with a red line at the collar and cuffs; or perhaps a white sailor suit would be pretty. And in winter a little fur coat, with a squirrelskin cap, and possibly Russian boots . . ." Her imagination was running away with her. "And I'd let his hair grow, like a page's, and have it just curled up a little at the tips. And a straight fringe across his forehead. Everyone would turn round and stare after us if I took him out with me in Via Tornabuoni."

What you want, I should have liked to tell her, is not a child; it's a clockwork doll or a performing monkey. But I did not say so—partly because I could not think of the Italian for a clockwork doll and partly because I did not want to risk having the rent raised another fifteen per cent.

"Ah, if I only had a little boy like that!" She sighed and modestly dropped her eyelids. "I adore children. I sometimes think of adopting one—that is, if my husband would allow it."

I thought of the poor old gentleman being dragged along at the heels of his big white dog and inwardly smiled.

"But I don't know if he would," the Signora was continuing, "I don't know if he would." She was silent for a moment, as though considering a new idea.

A few days later, when we were sitting in the garden after luncheon, drinking our coffee, Guido's father, instead of passing with a nod and the usual cheerful good day, halted in front of us and began to talk. He was a fine handsome man, not very tall, but well proportioned, quick and elastic in his movements, and full of life. He had a thin brown face, featured like a Roman's and lit by a pair of the most intelligent-looking gray eyes I ever saw. They exhibited almost too much intelligence when, as not infrequently happened, he was trying, with an assumption of perfect frankness and a childlike innocence, to take one in or get something out of one. Delighting in itself, the intelligence shone there mischievously. The face might be ingenuous, impassive, almost imbecile in its expression; but the eyes on these occasions gave him completely away. One knew, when they glittered like that, that one would have to be careful.

Today, however, there was no dangerous light in them. He wanted nothing out of us, nothing of any value—only advice, which is a commodity, he knew, that most people are only too happy to part with. But he wanted advice on what was, for us, rather a delicate subject: on Signora Bondi. Carlo had often complained to us about her. The old man is good, he told us, very good and kind indeed. Which meant, I dare say, among other things, that he could easily be swindled. But his wife. . . . Well, the woman was a beast. And he would tell us stories of her insatiable rapacity: she was always claiming more than the half of the produce which, by the laws of the *métayage* systems, was the proprietor's due. He complained of her suspiciousness: she was forever accusing him of sharp practices, of downright stealing—him, he struck his breast, the soul of honesty. He complained of her shortsighted avarice: she wouldn't spend enough on manure, wouldn't buy him another cow, wouldn't have electric light installed in the stables. And we had sympathized, but cautiously, without expressing too strong an opinion on the subject. The Italians are wonderfully noncommittal in their speech; they will give nothing away to an interested person until they are quite certain that it is right and necessary and, above all, safe to do so. We had lived long enough among them to imitate their caution. What we said to Carlo would be sure, sooner or later, to get back to Signora Bondi. There was nothing to be gained by unnecessarily embittering our relations with the lady—only another fifteen per cent, very likely, to be lost.

Today he wasn't so much complaining as feeling perplexed. The Signora had sent for him, it seemed, and asked him how he would like it if she were to make an offer—it was all very hypothetical in the cautious Italian style—to adopt little Guido. Carlo's first instinct had been to say that he wouldn't like it at all. But an answer like that would have been too coarsely committal. He had preferred to say that he would think about it. And now he was asking for our advice.

Do what you think best, was what in effect we replied. But we gave it distantly but distinctly to be understood that we didn't think that Signora Bondi would make a very good foster mother for the child. And Carlo was inclined to agree. Besides he was very fond of the boy.

"But the thing is," he concluded rather gloomingly, "that if she has really set her heart on getting hold of the child, there's nothing she won't do to get him—nothing."

He too, I could see, would have liked the physicists to start on unemployed childless women of sanguine temperament before they tried to tackle the atom. Still, I reflected, as I watched him striding away along the terrace, singing powerfully from a brazen gullet as he went, there was force there, there was life enough in those elastic limbs, behind those bright gray eyes, to put up a good fight even against the accumulated vital energies of Signora Bondi.

It was a few days after this that my gramophone and two or three boxes of records arrived from England. They were a great comfort to us on the hilltop, providing as they did the only thing in which that spiritually fertile solitude—otherwise a perfect Swiss Family Robinson's island—was lacking: music. There is not much music to be heard nowadays in Florence. The times when Dr. Burney could tour through Italy, listening to an unending succession of new operas, symphonies, quartets, cantatas, are gone. Gone are the days when a learned musician, inferior only to the Reverend Father Martini of Bologna, could admire what the peasants sang and the strolling players thrummed and scraped on their instruments. I have traveled for weeks through the peninsula and hardly heard a note that was not *Salome* or the Fascists' song. Rich in nothing else that makes life agreeable or even supportable, the northern metropolises are rich in music. That is perhaps the only inducement that a reasonable man can find for living there. The other attractions—organized gaiety, people, miscellaneous conversation, the social pleasures—what are those, after all, but an expense of spirit that buys nothing in return? And then the cold, the darkness, the moldering dirt, the damp and squalor. . . . No, where there is no necessity that retains, music can be the only inducement. And that, thanks to the ingenious Edison, can now be taken about in a box and unpacked in whatever solitude one chooses to visit. One can live at Benin, or Nuneaton, or Tozeur in the Sahara, and still hear Mozart quartets, and selections from *The Well-Tempered Clavichord*, and the Fifth Symphony, and the Brahms clarinet quintet, and motets by Palestrina.

Carlo, who had gone down to the station with his mule and cart to fetch the packing case, was vastly interested in the machine.

"One will hear some music again," he said, as he watched me unpacking the gramophone and the disks. "It is difficult to do much oneself."

Still, I reflected, he managed to do a good deal. On warm nights we used to hear him, where he sat at the door of his house, playing his guitar and softly singing; the eldest boy shrilled out the melody on the mandolin and sometimes the whole family would join in, and the darkness would be filled with their passionate, throaty singing. Piedigrotta songs they mostly sang; and the voices drooped slurringly from note to note, lazily climbed or jerked themselves with sudden sobbing emphases from one tone to another. At a distance and under the stars the effect was not unpleasing.

"Before the war," he went on, "in normal times" (and Carlo had a hope, even a belief, that the normal times were coming back and that life would soon be as cheap and easy as it had been in the days before the flood), "I used to go and listen to the operas at the Politeama. Ah, they were magnificent. But it costs five lire now to get in."

"Too much," I agreed.

"Have you got *Trovatore?*" he asked.

I shook my head.

"*Rigoletto?*"

"I'm afraid not."

"*Bohème? Fanciulla del West? Pagliacci?*"

I had to go on disappointing him.

"Not even *Norma?* Or the *Barbiere?*"

I put on Battistini in "*Là ci darem*" out of *Don Giovanni*. He agreed that the singing was good; but I could see that he didn't much like the music. Why not? He found it difficult to explain.

"It's not like *Pagliacci*," he said at last.

"Not palpitating?" I suggested, using a word with which I was sure he would be familiar; for it occurs in every Italian political speech and patriotic leading article.

"Not palpitating," he agreed.

And I reflected that it is precisely by the difference between *Pagliacci* and *Don Giovanni*, between the palpitating and the nonpalpitating, that modern music taste is separated from the old. The corruption of the best, I thought, is the worst. Beethoven taught music to palpitate with his intellectual and spiritual passion. It has gone on palpitating ever since, but with the passion of inferior men. Indirectly, I thought, Beethoven is responsible for *Parsifal*, *Pagliacci*, and the *Poem of Fire*; still more indirectly for *Samson and Delilah* and "Ivy, cling to me." Mozart's melodies may be brilliant, memorable, infectious; but they don't palpitate, don't catch you between wind and water, don't send the listener off into erotic ecstasies.

Carlo and his elder children found my gramophone, I am afraid, rather a disappointment. They were too polite, however, to say so openly; they merely ceased, after the first day or two, to take any interest in the machine and the music it played. They preferred the guitar and their own singing.

Guido, on the other hand, was immensely interested. And he liked, not the cheerful dance tunes, to whose sharp rhythms our little Robin loved to go stamping round and round the room, pretending that he was a whole regiment of soldiers, but the genuine stuff. The first record he heard, I remember, was that of the slow movement of Bach's Concerto in D Minor for two violins. That was the disk I put on the turntable as soon as Carlo had left me. It seemed to me, so to speak, the most musical piece of music with which I would refresh my long-parched mind—the coolest and clearest of all draughts. The movement had just got under way and was beginning to unfold its pure and melancholy beauties in accordance with the laws of the most exacting intellectual logic, when the two children, Guido in front and little Robin breathlessly following, came clattering into the room from the loggia.

Guido came to a halt in front of the gramophone and stood there, motionless, listening. His pale blue-gray eyes opened themselves wide; making a little nervous gesture that I had often noticed in him before, he plucked at his lower lip with his thumb and forefinger. He must have taken a deep breath; for I noticed that, after listening for a few seconds, he sharply expired and drew in a fresh gulp of air. For an instant he looked at me—a questioning, astonished, rapturous look—gave a little laugh that ended in a kind of nervous shudder, and turned back toward the source of the incredible sounds. Slavishly imitating his elder comrade, Robin had also taken up his stand in front of the gramophone, and in exactly the same position, glancing at Guido from time to time to make sure that he was doing everything, down to plucking at his lip, in the correct way. But after a minute or so he became bored.

"Soldiers," he said, turning to me; "I want soldiers. Like in London." He remembered the ragtime and the jolly marches round and round the room.

I put my fingers to my lips. "Afterwards," I whispered.

Robin managed to remain silent and still for perhaps another twenty seconds. Then he seized Guido by the arm, shouting, "*Vieni*, Guido! Soldiers. *Soldati. Vieni giuocare soldati.*"

It was then, for the first time, that I saw Guido impatient. "*Vai!*" he whispered angrily, slapped at Robin's clutching hand and pushed him roughly away. And he leaned a little closer to the instrument, as though to make up by yet intenser listening for what the interruption had caused him to miss.

Robin looked at him, astonished. Such a thing had never happened before. Then he burst out crying and came to me for consolation.

When the quarrel was made up—and Guido was sincerely repentant, was as nice as he knew how to be when the music had stopped and his mind was free to think of Robin once more—I asked him how he liked the music. He said he thought it was beautiful. But *bello* in Italian is too vague a word, too easily and frequently uttered, to mean very much.

"What did you like best?" I insisted. For he had seemed to enjoy it so much that I was curious to find out what had really impressed him.

He was silent for a moment, pensively frowning. "Well," he said at last, "I liked the bit that went like this." And he hummed a long phrase. "And then there's the other thing singing at the same time—but what are those things," he interrupted himself, "that sing like that?"

"They're called violins," I said.

"Violins." He nodded. "Well, the other violin goes like this." He hummed again. "Why can't one sing both at once? And what is in that box? What makes it make that noise?" The child poured out his questions.

I answered him as best I could, showing him the little spirals on the disk, the needle, the diaphragm. I told him to remember how the string of the guitar trembled when one plucked it; sound is a shaking in the air, I told him, and I tried to explain how those shakings get printed on the black disk. Guido listened to me very gravely, nodding from time to time. I had the impression that he understood perfectly well everything I was saying.

By this time, however, poor Robin was so dreadfully bored that in pity for him I had to send the two children out into the garden to play. Guido went obediently; but I could see that he would have preferred to stay indoors and listen to more music. A little while later, when I looked out, he was hiding in the dark recesses of the big bay tree, roaring like a lion, and Robin laughing, but a little nervously, as though he were afraid that the horrible noise might possibly turn out, after all, to be the roaring of a real lion, was beating the bush with a stick, and shouting, "Come out, come out! I want to shoot you."

After lunch, when Robin had gone upstairs for his afternoon sleep, he reappeared. "May I listen to music now?" he asked. And for an hour he sat there in front of the instrument, his head cocked slightly on one side, listening while I put on one disk after another.

Thenceforward he came every afternoon. Very soon he knew all my library of records, had his preferences and dislikes, and could ask for what he wanted by humming the principal theme.

"I don't like that one," he said of Strauss's *Till Eulenspiegel*. "It's like what we sing in our house. Not really like, you know. But somehow rather like, all the same. You understand?" He looked at us perplexedly and appealingly, as though begging us to understand what he meant and so save him from going on explaining. We nodded. Guido went on. "And then," he said, "the end doesn't seem to come properly out of the beginning. It's not like the one you played the first time." He hummed a bar or two from the slow movement of Bach's D Minor Concerto.

"It isn't," I suggested, "like saying: All little boys like playing. Guido is a little boy. Therefore Guido likes playing."

He frowned. "Yes, perhaps that's it," he said at last. "The one you played first is more like that. But, you know," he added, with an excessive regard for truth, "I don't like playing as much as Robin does."

Wagner was among his dislikes; so was Debussy. When I played the record of one of Debussy's arabesques, he said, "Why does he say the same thing over and over again? He ought to say something new, or go on, or make the thing grow. Can't he think of anything different?" But he was less censorious about the *Après-midi d'un faune*. "The things have beautiful voices," he said.

Mozart overwhelmed him with delight. The duet from *Don Giovanni,* which his father had found insufficiently palpitating, enchanted Guido. But he preferred the quartets and the orchestral pieces.

"I like music," he said, "better than singing."

Most people, I reflected, like singing better than music; are more interested in the executant than in what he executes, and find the impersonal orchestra less moving than the soloist. The touch of the pianist is the human touch, and the soprano's high C is the personal note. It is for the sake of this touch, that note, that audiences fill the concert halls.

Guido, however, preferred music. True, he liked "*Là ci darem*"; he liked "*Deh vieni alla finestra*"; he thought "*Che soave zefiretto*" so lovely that almost all our concerts had to begin with it. But he preferred the other things. The *Figaro* overture was one of his favorites. There is a passage not far from the beginning of the piece, when the first violins suddenly go rocketing up into the heights of loveliness; as the music approached that point, I used always to see a smile developing and gradually brightening on Guido's face, and when, punctually, the thing happened, he clapped his hands and laughed aloud with pleasure.

On the other side of the same disk, it happened, was recorded Beethoven's *Egmont* overture. He liked that almost better than *Figaro.*

"It has more voices," he explained. And I was delighted by the acuteness of the criticism; for it is precisely in the richness of its orchestration that *Egmont* goes beyond *Figaro*.

But what stirred him almost more than anything was the *Coriolan* overture. The third movement of the Fifth Symphony, the second movement of the Seventh, the slow movement of the Emperor Concerto—all these things ran it pretty close. But none excited him so much as *Coriolan*. One day he made me play it three or four times in succession; then he put it away.

"I don't think I want to hear that any more," he said.

"Why not?"

"It's too . . . too . . ." He hesitated. "Too big," he said at last. "I don't really understand it. Play me the one that goes like this." He hummed the phrase from the D Minor Concerto.

"Do you like that one better?" I asked.

He shook his head. "No, it's not that exactly. But it's easier."

"Easier?" It seemed to me rather a queer word to apply to Bach.

"I understand it better."

One afternoon, while we were in the middle of our concert, Signora Bondi was ushered in. She began at once to be overwhelmingly affectionate toward the child; kissed him, patted his head, paid him the most outrageous compliments on his appearance. Guido edged away from her.

"And do you like music?" she asked.

The child nodded.

"I think he has a gift," I said. "At any rate, he has a wonderful ear and a power of listening and criticizing such as I've never met with in a child of that age. We're thinking of hiring a piano for him to learn on."

A moment later I was cursing myself for my undue frankness in praising the boy. For Signora Bondi began immediately to protest that, if she could have the upbringing of the child, she would give him the best masters, bring out his talent, make an accomplished maestro of him—and, on the way, an infant prodigy. And at that moment, I am sure, she saw herself sitting maternally, in pearls and black satin, in the lea of the huge Steinway, while an angelic Guido, dressed like little Lord Fauntleroy, rattled out Liszt and Chopin to the loud delight of a thronged auditorium. She saw the bouquets and all the elaborate floral tributes, heard the clapping and the few well-chosen words with which the veteran maestri, touched almost to tears, would hail the coming of the little genius. It became more than ever important for her to acquire the child.

"You've sent her away fairly ravening," said Elizabeth, when Signora Bondi had gone. "Better tell her next time that you made a mistake, and that the boy's got no musical talent whatever."

In due course a piano arrived. After giving him the minimum of preliminary instruction, I let Guido loose on it. He began by picking out for himself the melodies he had heard, reconstructing the harmonies in which they were embedded. After a few lessons, he understood the rudiments of musical notation and could read a simple passage at sight, albeit very slowly. The whole process of reading was still strange to him; he had picked up his letters somehow, but nobody had yet taught him to read whole words and sentences.

I took occasion, next time I saw Signora Bondi, to assure her that Guido had disappointed me. There was nothing in his musical talent, really. She professed to be very sorry to hear it; but I could see that she didn't for a moment believe me. Probably she thought that we were after the child too, and wanted to bag the infant prodigy for ourselves, before she could get in her claim, thus depriving her of what she regarded almost as her feudal right. For, after all, weren't they her peasants? If anyone was to profit by adopting the child it ought to be herself.

Tactfully, diplomatically, she renewed her negotiations with Carlo. The boy, she put it to him, had genius. It was the foreign gentleman who had told her so, and he was the sort of man, clearly, who knew about such things. If Carlo would let her adopt the child, she'd have him trained. He'd become a great maestro and get engagements in the Argentine and the United States, in Paris and London. He'd earn millions and millions. Think of Caruso, for example. Part of the millions, she explained, would of course come to Carlo. But before they began to roll in, those millions, the boy would have to be trained. But training was very expensive. In his own interest, as well as that of his son, he ought to let her take charge of the child. Carlo said he would think it over, and again applied to us for advice. We suggested that it would be best in any case to wait a little and see what progress the boy made.

He made, in spite of my assertions to Signora Bondi, excellent progress. Every afternoon, while Robin was asleep, he came for his concert and his lesson. He was getting along famously with his reading; his small fingers were acquiring strength and agility. But what to me was more interesting was that he had begun to make up little pieces on his own account. A few of them I took down as he played them and I have them still. Most of them, strangely enough, as I thought then, are canons. He had a passion for canons. When I explained to him the principles of the form he was enchanted.

"It is beautiful," he said, with admiration. "Beautiful, beautiful. And so easy!"

Again the word surprised me. The canon is not, after all, so conspicuously simple. Thenceforward he spent most of his time at the piano in working out little canons for his own amusement. They were often remarkably ingenious. But in the invention of other kinds of music he did not show himself so fertile as I had hoped. He composed and harmonized one or two solemn little airs like hymn tunes, with a few sprightlier pieces in the spirit of the military march. They were extraordinary, of course, as being the inventions of a child. But a great many children can do extraordinary things; we are all geniuses up to the age of ten. But I had hoped that Guido was a child who was going to be a genius at forty; in which case what was extraordinary for an ordinary child was not extraordinary enough for him. "He's hardly a Mozart," we agreed, as we played his little pieces over. I felt, it must be confessed, almost aggrieved. Anything less than a Mozart, it seemed to me, was hardly worth thinking about.

He was not a Mozart. No. But he was somebody, as I was to find out, quite as extraordinary. It was one morning in the early summer that I made the discovery. I was sitting in the warm shade of our westward-facing balcony, working. Guido and Robin were playing in the little enclosed garden below. Absorbed in my work, it was only, I suppose, after the silence had prolonged itself a considerable time that I became aware that the children were making remarkably little noise. There was no shouting, no running about; only a quiet talking. Knowing by experience that when children are quiet it generally means that they are absorbed in some delicious mischief, I got up from my chair and looked over the balustrade to see what they were doing. I expected to catch them dabbling in water, making a bonfire, covering themselves with tar. But what I actually saw was Guido, with a burnt stick in his hand, demonstrating on the smooth paving stones of the path, that the square on the hypotenuse of a right-angled triangle is equal to the sum of the squares on the other two sides.

Kneeling on the floor, he was drawing with the point of his blackened stick on the flagstones. And Robin, kneeling imitatively beside him, was growing, I could see, rather impatient with this very slow game.

"Guido," he said. But Guido paid no attention. Pensively frowning, he went on with his diagram. "Guido!" The younger child bent down and then craned round his neck so as to look up into Guido's face. "Why don't you draw a train?"

"Afterwards," said Guido. "But I just want to show you this first. It's so beautiful," he added cajolingly.

"But I want a train," Robin persisted.

"In a moment. Do just wait a moment." The tone was almost imploring. Robin armed himself with renewed patience. A minute later Guido had finished both his diagrams.

"There!" he said triumphantly, and straightened himself up to look at them. "Now I'll explain."

And he proceeded to prove the theorem of Pythagoras—not in Euclid's way, but by the simpler and more satisfying method which was, in all probability, employed by Pythagoras himself. He had drawn a square and dissected it, by a pair of crossed perpendiculars, into two squares and two equal rectangles. The equal rectangles he divided up by their diagonals into four equal right-angled triangles. The two squares are then seen to be the squares on the two sides of any one of these triangles other than the hypotenuse. So much for the first diagram. In the next he took the four right-angled triangles into which the rectangles had been divided and rearranged them round the original square so that their right angles filled the corners of the square, the hypotenuses looked inward, and the greater and less sides of the triangles were in continuation along the sides of the squares (which are each equal to the sum of these sides). In this way the original square is redissected into four right-angled triangles and the square on the hypotenuse. The four triangles are equal to the two rectangles of the original dissection. Therefore the square on the hypotenuse is equal to the sum of the two squares—the squares on the two other sides—

into which, with the rectangles, the original square was first dissected.

In very untechnical language, but clearly and with a relentless logic, Guido expounded his proof. Robin listened, with an expression on his bright, freckled face of perfect incomprehension.

"*Treno*," he repeated from time to time. "*Treno*. Make a train."

"In a moment," Guido implored. "Wait a moment. But do just look at this. Do." He coaxed and cajoled. "It's so beautiful. It's so easy."

So easy . . . The theorem of Pythagoras seemed to explain for me Guido's musical predilections. It was not an infant Mozart we had been cherishing; it was a little Archimedes with, like most of his kind, an incidental musical twist.

"*Treno, treno!*" shouted Robin, growing more and more restless as the exposition went on. And when Guido insisted on going on with his proof, he lost his temper. "*Cattivo* Guido," he shouted, and began to hit out at him with his fists.

"All right," said Guido resignedly. "I'll make a train." And with his stick of charcoal he began to scribble on the stones.

I looked on for a moment in silence. It was not a very good train. Guido might be able to invent for himself and prove the theorem of Pythagoras: but he was not much of a draftsman.

"Guido!" I called. The two children turned and looked up. "Who taught you to draw those squares?" It was conceivable, of course, that somebody might have taught him.

"Nobody." He shook his head. Then, rather anxiously, as though he were afraid there might be something wrong about drawing squares, he went on to apologize and explain. "You see," he said, "it seemed to me so beautiful. Because those squares—" he pointed at the two small squares in the first figure—"are just as big as this one." And, indicating the square on the hypotenuse in the second diagram, he looked up at me with a deprecating smile.

I nodded. "Yes, it's very beautiful," I said—"it's very beautiful indeed."

An expression of delighted relief appeared on his face; he laughed with pleasure. "You see it's like this," he went on, eager to initiate me into the glorious secret he had discovered. "You cut these two long squares—" he meant the rectangles—"into two slices. And then there are four slices, all just the same, because, because— Oh, I ought to have said that before—because these long squares are the same because those lines, you see . . ."

"But I want a train," protested Robin.

Leaning on the rail of the balcony, I watched the children below. I thought of the extraordinary thing I had just seen and of what it meant.

I thought of the vast differences between human beings. We classify men by the color of their eyes and hair, the shape of their skulls. Would it not be more sensible to divide them up into intellectual species? There would be even wider gulfs between the extreme mental types than between a Bushman and a Scandinavian. This child, I thought, when he grows up, will be to me, intellectually, what a man is to his dog. And there are other men and women who are, perhaps, almost as dogs to me.

Perhaps the men of genius are the only true men. In all the history of the race there have been only a few thousand real men. And the rest of us—what are we? Teachable animals. Without the help of the real man, we should have found out almost nothing at all. Almost all the ideas with which we are familiar could never have occurred to minds like ours. Plant the seeds there and they will grow; but our minds could never spontaneously have generated them.

There have been whole nations of dogs, I thought; whole epochs in which no Man was born. From the dull Egyptians the Greeks took crude experience and rules of thumb and made sciences. More than a thousand years passed before Archimedes had a comparable successor. There has been only one Buddha, one Jesus, only one Bach that we know of, one Michelangelo.

Is it by a mere chance, I wondered, that a Man is born from time to time? What causes a whole constellation of them to come contemporaneously into being and from out of a single people? Taine thought that Leonardo, Michelangelo, and Raphael were born when they were because the time was ripe for great painters and the Italian scene congenial. In the mouth of a rationalizing nineteenth-century Frenchman the doctrine is strangely mystical; it may be none the less true for that. But what of those born out of time? Blake, for example. What of those?

This child, I thought, has had the fortune to be born at a time when he will be able to make good use of his capacities. He will find the most elaborate analytical methods lying ready to his hand; he will have a prodigious experience behind him. Suppose him born while Stonehenge was building; he might have spent a lifetime discovering the rudiments, guessing darkly where now he might have had a chance of proving. Born at the time of the Norman Conquest, he would have had to wrestle with all the preliminary difficulties created by an inadequate symbolism; it would have taken him long years, for example, to learn the art of dividing MMMCCCCLXXXVIII by MCMXIX. In five years, nowadays, he will learn what it took generations of Men to discover.

And I thought of the fate of all the Men born so hopelessly out of time that they could achieve little or nothing of value. Beethoven born in Greece, I thought, would have had to be content to play thin melodies on the flute or lyre; in those intellectual surroundings it would hardly have been possible for him to imagine the nature of harmony.

From drawing trains, the children in the garden below had gone onto playing trains. They were trotting round and round; with blown round cheeks and pouting mouth like the cherubic symbol of a wind, Robin puffed-puffed and Guido, holding the skirt of his smock, shuffled behind him, tooting. They ran forward, backed, stopped at imaginary stations, shunted, roared over bridges, crashed through tunnels, met with occasional collisions and derailments. The young Archimedes seemed to be just as happy as the little towheaded barbarian. A few minutes ago he had been busy with the theorem of Pythagoras. Now, tooting indefatigably along imaginary rails, he was perfectly content to shuffle backward and forward among the flower beds, between the pillars of the loggia, in and out of the dark tunnels of the laurel tree. The fact that one is going to be Archimedes does not prevent one from being an ordinary cheerful child meanwhile. I thought of this strange talent distinct and separate from the rest of the mind, independent, almost, of experience. The typical child prodigies are musical and mathematical; the other talents ripen slowly under the influence of emotional experience and growth. Till he was thirty Balzac gave proof of nothing but ineptitude; but at four the young Mozart was already a musician, and some of Pascal's most brilliant work was done before he was out of his teens.

In the weeks that followed, I alternated the daily piano lessons with lessons in mathematics. Hints rather than lessons they were; for I only made suggestions, indicated methods, and left the child himself to work out the ideas in detail. Thus I introduced him to algebra by showing him another proof of the theorem of Pythagoras. In this proof one drops a perpendicular from the right angle on to the hypotenuse, and arguing from the fact that the two triangles thus created are similar to one another and to the original triangle, and that the proportions which their corresponding sides bear to one another are therefore equal, one can show in algebraical form that $C^2 + D^2$ (the squares on the other two sides) are equal to $A^2 + B^2$ (the squares on the two segments of the hypotenuses) + 2AB; which last, it is easy to show geometrically, is equal to $(A + B)^2$, or the square on the hypotenuse. Guido was as much enchanted by the rudiments of algebra as he would have been if I had given him an engine worked by steam, with a methylated spirit lamp to heat the boiler; more enchanted, perhaps—for the engine would have got broken, and, remaining always itself, would in any case have lost its charm, while the rudiments of algebra continued to grow and blossom in his mind with an unfailing luxuriance. Every day he made the discovery of something which seemed to him exquisitely beautiful; the new toy was inexhaustible in its potentialities.

In the intervals of applying algebra to the second book of Euclid, we experimented with circles; we stuck bamboos into the parched earth, measured their shadows at different hours of the day, and drew exciting conclusions from our observations. Sometimes, for fun, we cut and folded sheets of paper so as to make cubes and pyramids. One afternoon Guido arrived carrying carefully between his small and rather grubby hands a flimsy dodecahedron.

"*E tanto bello!*" he said, as he showed us his paper crystal; and when I asked him how he had managed to make it, he merely smiled and said it had been so easy. I looked at Elizabeth and laughed. But it would have been more symbolically to the point, I felt, if I had gone down on all fours, wagged the spiritual outgrowth of my os coccyx, and barked my astonished admiration.

It was an uncommonly hot summer. By the beginning of July our little Robin, unaccustomed to these high temperatures, began to look pale and tired; he was listless, had lost his appetite and energy. The doctor advised mountain air. We decided to spend the next ten or twelve weeks in Switzerland. My parting gift to Guido was the first six books of Euclid in Italian. He turned over the pages, looking ecstatically at the figures.

"If only I knew how to read properly," he said. "I'm so stupid. But now I shall really try to learn."

From our hotel near Grindelwald we sent the child, in Robin's name, various postcards of cows, alphorns, Swiss chalets, edelweiss, and the like. We received no answers to these cards; but then we did not expect answers. Guido could not write, and there was no reason why his father or his sisters should take the trouble to write for him. No news, we took it, was good news. And then one day, early in September, there arrived at the hotel a strange letter. The manager had it stuck up on the glass-fronted notice board in the hall, so that all the guests might see it, and whoever conscientiously thought that it belonged to him might claim it. Passing the board on the way in to lunch, Elizabeth stopped to look at it.

"But it must be from Guido," she said.

I came and looked at the envelope over her shoulder. It was unstamped and black with postmarks. Traced out in pencil, the big uncertain capital letters sprawled across its face. In the first line was written: AL BABBO DI ROBIN, and there followed a travestied version of the name of the hotel and the place. Round the address bewildered postal officials had scrawled suggested emendations. The letter had wandered for a fortnight at least, back and forth across the face of Europe.

"*Al Babbo di Robin.* To Robin's father." I laughed. "Pretty smart of the postmen to have got it here at all." I went to the manager's office, set forth the justice of my claim to the letter and, having paid the fifty-centime surcharge for the missing stamp, had the case unlocked and the letter given me. We went in to lunch.

"The writing's magnificent," we agreed, laughing, as we examined the address at close quarters. "Thanks to Euclid," I added. "That's what comes of pandering to the ruling passion."

But when I opened the envelope and looked at its contents I no longer laughed. The letter was brief and almost telegraphic in style. SONO DALLA PADRONA, it ran, NOM MI PIACE HA RUBATO IL MIO LIBRO NON VOGLIO SUONARE PIU VOGLIO TORNARE A CASA VENGA SUBITO GUIDO.

"What is it?"

I handed Elizabeth the letter. "That blasted woman's got hold of him," I said.

Busts of men in Homburg hats, angels bathed in marble tears extinguishing torches, statues of little girls, cherubs, veiled figures, allegories and ruthless realisms—the strangest and most diverse idols beckoned and gesticulated as we passed. Printed indelibly on tin and embedded in the living rock, the brown photographs looked out, under glass, from the humbler crosses, headstones, and broken pillars. Dead ladies in cubistic geometrical fashions of thirty years ago—two cones of black satin meeting point to point at the waist, and the arms: a sphere to the elbow, a polished cylinder below—smiled mournfully out of their marble frames; the smiling frames; the smiling faces, the white hands, were the only recognizably human things that emerged from the solid geometry of their clothes. Men with black mustaches, men with white beards, young clean-shaven men, stared or averted their gaze to show a Roman profile. Children in their stiff best opened wide their eyes, smiled hopefully in anticipation of the little bird that was to issue from the camera's muzzle, smiled skeptically in the knowledge that it wouldn't, smiled laboriously and obediently because they had been told to. In spiky Gothic cottages of marble the richer dead reposed; through grilled doors one caught a glimpse of pale Inconsolables weeping, of distraught Geniuses guarding the secret of the tomb. The less prosperous sections of the majority slept in communities, close-crowded but elegantly housed under smooth continuous marble floors, whose every flagstone was the mouth of a separate grave.

These Continental cemeteries, I thought, as Carlo and I made our way among the dead, are more frightful than ours, because these people pay more attention to their dead than we do. That primordial cult of corpses, that tender solicitude for their material well-being, which led the ancients to house their dead in stone, while they themselves lived between wattles and under thatch, still lingers here; persists, I thought, more vigorously than with us. There are a hundred gesticulating statues here for every one in an English graveyard. There are more family vaults, more "luxuriously appointed" (as they say of liners and hotels) than one would find at home. And embedded in every tombstone there are photographs to remind the powdered bones within what form they will have to resume on the Day of Judgment;

beside each are little hanging lamps to burn optimistically on All Souls' Day. To the Man who built the Pyramids they are nearer, I thought, than we.

"If I had known," Carlo kept repeating, "if only I had known." His voice came to me through my reflections as though from a distance. "At the time he didn't mind at all. How should I have known that he would take it so much to heart afterwards? And she deceived me, she lied to me."

I assured him yet once more that it wasn't his fault. Though, of course, it was, in part. It was mine too, in part; I ought to have thought of the possibility and somehow guarded against it. And he shouldn't have let the child go, even temporarily and on trial, even though the woman was bringing pressure to bear on him. And the pressure had been considerable. They had worked on the same holding for more than a hundred years, the men of Carlo's family; and now she had made the old man threaten to turn him out. It would be a dreadful thing to leave the place; and besides, another place wasn't so easy to find. It was made quite plain, however, that he could stay if he let her have the child. Only for a little to begin with; just to see how he got on. There would be no compulsion whatever on him to stay if he didn't like it. And it would be all to Guido's advantage; and to his father's, too, in the end. All that the Englishman had said about his not being such a good musician as he had thought at first was obviously untrue—mere jealousy and little-mindedness: the man wanted to take credit for Guido himself, that was all. And the boy, it was obvious, would learn nothing from him. What he needed was a real good professional master.

All the energy that, if the physicists had known their business, would have been driving dynamos, went into this campaign. It began the moment we were out of the house, intensively. She would have more chance of success, the Signora doubtless thought, if we weren't there. And besides, it was essential to take the opportunity when it offered itself and get hold of the child before we could make our bid—for it was obvious to her that we wanted Guido just as much as she did.

Day after day she renewed the assault. At the end of a week she sent her husband to complain about the state of the vines: they were in a shocking condition; he had decided, or very nearly decided, to give Carlo notice. Meekly, shamefacedly, in obedience to higher orders, the old gentleman uttered his threats. Next day Signora Bondi returned to the attack. The padrone, she declared, had been in a towering passion; but she'd do her best, her very best, to mollify him. And after a significant pause she went on to talk about Guido.

In the end Carlo gave in. The woman was too persistent and she held too many trump cards. The child could go and stay with her for a month or two on trial. After that, if he really expressed a desire to remain with her, she would formally adopt him.

At the idea of going for a holiday to the seaside—and it was to the seaside, Signora Bondi told him, that they were going—Guido was pleased and excited. He had heard a lot about the sea from Robin. "*Tanta acqua!*" It had sounded almost too good to be true. And now he was actually to go and see this marvel. It was very cheerfully that he parted from his family.

But after the holiday by the sea was over, and Signora Bondi had brought him back to her town house in Florence, he began to be homesick. The Signora, it was true, treated him exceedingly kindly, bought him new clothes, took him out to tea in the Via Tornabuoni and filled him up with cakes, iced strawberry-ade, whipped cream, and chocolates. But she made him practice the piano more than he liked, and what was worse, she took away his Euclid, on the score that he wasted too much time with it. And when he said that he wanted to go home, she put him off with promises and excuses and downright lies. She told him that she couldn't take him at once, but that next week, if he were good and worked hard at his piano meanwhile, next week . . . And when the time came she told him that his father didn't want him back. And she redoubled her petting, gave him expensive presents, and stuffed him with yet unhealthier foods. To no purpose. Guido didn't like his new life, didn't want to practice scales, pined for his book, and longed to be back with his brothers and sisters. Signora Bondi, meanwhile, continued to hope that time and chocolates would eventually make the child hers; and to keep his family at a distance, she wrote to Carlo every few days letters which still purported to come from the seaside (she took the trouble to send them to a friend, who posted them back again to Florence), and in which she painted the most charming picture of Guido's happiness.

It was then that Guido wrote his letter to me. Abandoned, as he supposed, by his family—for that they should not take the trouble to come to see him when they were so near was only to be explained on the hypothesis that they really had given him up—he must have looked to me as his last and only hope. And the letter, with its fantastic address, had been nearly a fortnight on its way. A fortnight—it must have seemed hundreds of years; and as the centuries succeeded one another, gradually, no doubt, the poor child became convinced that I too had abandoned him. There was no hope left.

"Here we are," said Carlo.

I looked up and found myself confronted by an enormous monument. In a kind of grotto hollowed in the flanks of a monolith of gray sandstone, Sacred Love, in bronze, was embracing a funeral urn. And in bronze letters riveted into the stone was a long legend to the effect that the inconsolable Ernesto Bondi had raised this monument to the memory of his beloved wife, Anunziata, as a token of his undying love for one whom, snatched from him by a premature death, he hoped very soon to join beneath this stone. The first Signora Bondi had died in 1912. I thought of the old man leashed to his white dog; he must always, I reflected, have been a most uxorious husband.

"They buried him here."

We stood there for a long time in silence. I felt the tears coming into my eyes as I thought of the poor child lying there underground. I thought of those luminous grave eyes, and the curve of that beautiful forehead, the droop of the melancholy mouth, of the expression of delight which illumined his face when he learned of some new idea that pleased him, when he heard a piece of music that he liked. And this beautiful small being was dead; and the spirit that inhabited this form, the amazing spirit, that too had been destroyed almost before it had begun to exist.

And the unhappiness that must have preceded the final act, the child's despair, the conviction of his utter abandonment—those were terrible to think of, terrible.

"I think we had better come away now," I said at last, and touched Carlo on the arm. He was standing there like a blind man, his eyes shut, his face slightly lifted toward the light; from between his closed eyelids the tears welled out, hung for a moment, and trickled down his cheeks. His lips trembled and I could see that he was making an effort to keep them still. "Come away," I repeated.

The face which had been still in its sorrow was suddenly convulsed;

he opened his eyes, and through the tears they were bright with a violent anger. "I shall kill her," he said, "I shall kill her. When I think of him throwing himself out, falling through the air . . ." With his two hands he made a violent gesture, bringing them down from over his head and arresting them with a sudden jerk when they were on the level with his breast. "And then crash." He shuddered. "She's as much responsible as though she had pushed him down herself. I shall kill her." He clenched his teeth.

To be angry is easier than to be sad, less painful. It is comforting to think of revenge. "Don't talk like that," I said. "It's no good. It's stupid. And what would be the point?" He had had those fits before, when grief became too painful and he had tried to escape from it. Anger had been the easiest way of escape. I had had, before this, to persuade him back into the harder path of grief. "It's stupid to talk like that," I repeated, and I led him away through the ghastly labyrinth of tombs, where death seemed more terrible even than it is.

By the time we had left the cemetery, and were walking down from San Miniato toward the Piazzale Michelangelo below, he had become calmer. His anger had subsided again into the sorrow from which it had derived all its strength and its bitterness. In the Piazzale we halted for a moment to look down at the city in the valley below us. It was a day of floating clouds—great shapes, white, golden, and gray; and between them patches of a thin, transparent blue. Its lantern level, almost, with our eyes, the dome of the cathedral revealed itself in all its grandiose lightness, its vastness and aerial strength. On the innumerable brown and rosy roofs of the city the afternoon sunlight lay softly, sumptuously, and the towers were as though varnished and enameled with an old gold. I thought of all the Men who had lived here and left the visible traces of their spirit and conceived extraordinary things. I thought of the dead child.

ARTHUR KOESTLER

—

# Pythagoras and the Psychoanalyst

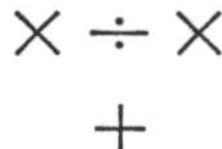

HE SPENT most of the time dozing on his bed, daydreaming for hours on end. At night he could not sleep, and his thoughts began to spin like a top which, when hit by the lash, jumps and changes its inclination but goes on spinning at the same speed; hits the walls, jumps, and goes on spinning at yet another angle. To stop this dizzy torment and to pass the time, he went on writing the story he had begun some days ago.

It was about a young man, sitting on a beach and drawing triangles with his stick in the sand. . . .

. . . He did not see the sea-gulls circling above his head nor the galleys and triremes moving softly along the water's skyline; he wore a strange, loose gown and his face was set in a dumb, anguished gaze on his figures in the sand, while his lips mumbled unintelligible words. An old man with shrewd, wrinkled eyes sat down at the other end of the bench; and after watching for a while the young man's antics, spoke to him in a gentle voice:

"What, my friend, are you doing with your stick?"

The young man jumped as if caught at a shameful or criminal occupation. "I am drawing triangles," he said, blushing foolishly.

"And why, after having drawn one, do you wipe it out with your hand, and then draw a new one which is just like the other?"

"I don't know. I believe these triangles have a secret, and I want to discover it."

[FROM *Arrival and Departure*, COURTESY THE MACMILLAN COMPANY AND JONATHAN CAPE LTD.]

"A secret? Tell me, my friend, do you perchance suffer from bad dreams? Do you cry out sometimes in your sleep?"

"I do, ever and anon."

"And what is the dream that haunts you and makes you cry out in the night?"

The young man blushed once more all over his face.

"I always dream that I and my dear wife Celia are watching the athletic games where my friend Porphyrius is performing; he throws the discus, but in the wrong direction, and the heavy thing comes whirling through the air and hits my poor wife on the head, who thereupon faints with a mysterious smile on her lips. . . ."

The old man chuckled and laid his hand on the other's shoulder.

"My dear friend," he said, "you are lucky that fate made me cross your path, for I am a teller of oracles, a solver of riddles, a helper of the afflicted. This will cost you a drachma, but it will be worth it. And now listen:

"I have noticed that while you were telling your dream, your hand again inadvertently began drawing in the sand. When you mentioned yourself, you drew a straight line. When you mentioned friend Porphyrius, you drew a second one at right angles to the first; and when you mentioned your wife Celia, you completed the triangle by drawing the hypotenuse which connects the other two. Thus your dream becomes perfectly transparent. Your mind is harassed by a disquietude which you have been hiding, even from yourself; and the secret of the triangle you are trying to discover can easily be solved by questioning your servants about your wife's private life."

The young man, whose name was Pythagoras, jumped to his feet. "Praised be the gods that you have solved the riddle which haunted my mind! Instead of going on drawing those foolish triangles, as I have done for the past two years, I shall now go home and give Celia a sound thrashing, as befits a reasonable man."

He stamped with his sandals on the last figure he had drawn, then, gathering up his robe, walked away with hurried steps along the beach. He felt happy and relieved; that dark, inexplicable urge to draw triangles in the sand had left him for ever; and thus the Pythagorean Proposition was never found.

RICHARD LLEWELLYN

# Mother and the Decimal Point

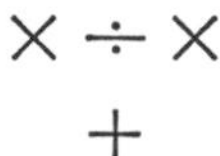

THAT NIGHT Mrs. Tom Jenkins came up to give me a polish in sums, written and mental. My father and mother, Ivor and Bron, and Davy were all round the table listening, and everybody quiet, pretending not to look.

We were doing very well, up to the kind of sum when a bath is filling at the rate of so many gallons and two holes are letting the water out, and please to say how long will it take to fill the bath, when my mother put down the socks she was darning and clicked her tongue in impatience.

"What is the matter?" my father asked her.

"That old National School," my mother said. "There is silly the sums are with them. Filling up an old bath with holes in it, indeed. Who would be such a fool?"

"A sum it is, girl," my father said. "A sum. A problem for the mind. Nothing to do with the National School, either."

"Filling the boy with old nonsense," Mama said.

"Not nonsense, Beth," my father said, to soothe, quietly, "a sum, it is. The water pours in and takes so long. It pours out and takes so long. How long to fill? That is all."

"But who would pour water in an old bath with holes?" my mother said. "Who would think to do it, but a lunatic?"

"Well, devil fly off," my father said, and put down his book to look at the ceiling. "It is to see if the boy can calculate, girl. Figures, nothing else. How many gallons and how long."

[FROM *How Green Was My Valley*. REPRINTED BY PERMISSION OF RICHARD LLEWELLYN AND THE MACMILLAN COMPANY]

"In a bath full of holes," Mama said, and rolled the sock in a ball and threw it in the basket, and it fell out, and she threw it back in twice as hard. "If he went to school in trews full of holes, we should hear about it. But an old bath can be so full with holes as a sieve and nobody taking notice."

"Look you," my father said to Mrs. Jenkins, "no more baths. Have you got something else?"

"Decimals, Mr. Morgan," said Mrs. Tom, "but he is strong in those."

"Decimals," said my father, "and peace in my house, for the love of God."

"Hisht," Mama said.

Decimals, then. And the look on my mother's face when the decimal point started his travels up and down the line was something to see.

In bed that night I heard my mother come upstairs and speak to Angharad, and then my father came up with the lamp, and left their door open a bit to hear the clock.

"Gwil," my mother said, "who is in charge of this decimal point?"

"Who?" my father said, and flap went his braces on the cupboard door.

"Decimal point," my mother said, "this thing Huw has got downstairs."

"More of this again, now," my father said, and laugh strong in his voice. "Look, Beth, my little one, leave it, now. Or else it will be morning and us fit for Bedlam, both."

"But what is it?" my mother said. "Why is a small boy allowed to know and I am such a fool?"

"Beth, Beth, Beth," my father said, "bless your sweet face, there are things for boys and things for girls. Decimal point makes fractions out of a whole. Instead of saying one and a half, you say one point five. Because five is half of ten, a one and a nought. The one is a whole one and nought is nothing. Now you are wiser."

Minutes went, and only the sound of clothes coming off and somebody late walking up the hill outside.

"But whose is it?" my mother said, as though a gate had been loosed. "Does it belong to somebody?"

"Well, Beth," my father said, "there is silly. Why should it belong to somebody? It is a decimal point, a dot on the paper. How can an ink dot belong to somebody?"

"Then who knows what is to be done with it?" my mother asked. "Multiply by ten, move the point, add a nought."

"No, girl," Dada said, "not add a nought. That is division. Multiplying, move the point down. Dividing, move the point up."

"Go on with you," Mama said, "it can stop where it is. I would like to know who found it out, anyhow."

"The French, I think," my father said, "and leave it now, will you?"

"Well, no wonder," my mother said, and glad to blame someone, see. "Those old Frenchies, is it? If I had known that, the book would never have come inside the house."

"O, Beth," said Dada, "there is an old beauty you are. Go now, before I will push you on the floor."

"Frenchie, indeed," my mother said, "and decimal points, move up and down. Like monkeys. With Frenchies and old baths full of holes, what will come to the boy?"

"A scholarship," said my father, "that is what I would like."

"Scholarship? Well, I hope so, indeed," my mother said, for the sound of the word was like the name of an anthem. "What the world is coming to, I cannot tell you."

"Sleep, now then," my father said, "not for you to worry about the world, is it? Think of the old Queen with a Jubilee of worry to think about, and be thankful."

"I wonder does she know about this decimal point?" my mother said.

"Oh, hell open and crack," said my father, and out went the lamp. "The poor old lady is asleep these hours. Let us follow. Good night, now."

"Go and scratch," said Mama.

## JAMES BRANCH CABELL

—

# Jurgen Proves It by Mathematics

$\times \div \times$
$+$

So THE FIELDS were captured by the Philistines, and Chloris and Jurgen and all the People of the Field were judged summarily. They were declared to be obsolete illusions, whose merited doom was to be relegated to limbo. To Jurgen this appeared unreasonable.

"For I am no illusion," he asserted. "I am manifestly flesh and blood, and in addition, I am the high King of Eubonia, and no less. Why, in disputing these facts you contest circumstances that are so well known hereabouts as to rank among mathematical certainties. And that makes you look foolish, as I tell you for your own good."

This vexed the leaders of the Philistines, as it always vexes people to be told anything for their own good. "We would have you know," said they, "that we are not mathematicians; and that, moreover, we have no kings in Philistia, where all must do what seems to be expected of them, and have no other law."

"How then, if you have no kings nor any code of law, can you or any other persons be the leaders of Philistia?"

"Why, but in Philistia, as elsewhere, it is expected that women and priests should behave unaccountably. Therefore all we who are women or priests do what we will in Philistia, and the men there obey us. And it is we, the priests of Philistia, who do not think you can possibly have any flesh and blood under a shirt which we recognize to be a conventional figure of speech. It does not stand to reason. And certainly you could not ever prove such a thing by mathematics; and to say so is nonsense."

"But I can prove it by mathematics, quite irrefutably. I can prove anything you require of me by whatever means you may prefer," said Jurgen, "for the simple reason that I am a monstrous clever fellow."

Then spoke the wise Queen Dolores, saying, "I have studied mathematics. I will question this young man, in my tent tonight, and in the morning I will report the truth as to his pretensions. Are you content to endure this interrogatory, my spruce young fellow who wear the shirt of a king?"

Jurgen looked full upon her; she was lovely as a hawk is lovely; and of all that Jurgen saw Jurgen approved. He assumed the rest to be in keeping, and deduced that Dolores was a fine woman.

"Madame and Queen," said Jurgen, "I am content. And I can promise to deal fairly with you."

So that evening Jurgen was conducted into the purple tent of Queen Dolores of Philistia. It was quite dark there, and Jurgen went in alone, and wondering what would happen next; but this scented darkness he found of excellent augury, if only because it prevented his shadow from following Jurgen into this quiet place.

"Now, you who claim to be flesh and blood, and as King of Eubonia, too," says the voice of Queen Dolores, "what is this nonsense you are talking about proving any such claims by mathematics?"

"Well, but my mathematics," replied Jurgen, "are Praxagorean."

"What, do you mean Praxagoras of Cos?"

"As if," scoffed Jurgen, "anybody had ever heard of any other Praxagoras!"

"But he, as I recall, belonged to the medical school of the Dogmatici," observed the wise Queen Dolores, "and was particularly celebrated for his researches in anatomy. Was he, then, also a mathematician?"

"The two are not incongruous, madame, as I would be delighted to demonstrate."

"Oh, nobody said that! For, indeed, it does seem to me I have heard of this Praxagorean system of mathematics, though, I confess, I have never studied it."

"Our school, madame, postulates, first of all, that since the science of mathematics is an abstract science, it is best inculcated by some concrete example."

Said the Queen, "But that sounds rather complicated."

"It occasionally leads to complications," Jurgen admitted, "through a choice of the wrong example. But the axiom is no less true."

"Come, then, and sit next to me on this couch if you can find it in the dark; and do you explain to me what you mean."

"Why, madame, by a concrete example I mean one that is perceptible to any of the senses—as to sight or hearing, or touch—"

"Oh, oh!" said the Queen, "now I perceive what you mean by a concrete example. And grasping this, I can understand that complications must of course arise from a choice of the wrong example."

"Well, then, madame, it is first necessary to implant in you, by the force of example, a lively sense of the peculiar character, and virtues and properties, of each of the numbers upon which is based the whole science of Praxagorean mathematics. For in order to convince you thoroughly, we must start far down, at the beginning of all things."

"I see," said the Queen, "or rather, in this darkness I cannot see at all, but I perceive your point. Your opening interests me: and you may go on."

"Now ONE, or the monad," says Jurgen, "is the principle and the end of all: it reveals the sublime knot which binds together the chain of causes; it is the symbol of identity, of equality, of existence, of conservation, and of general harmony." And Jurgen emphasized these characteristics vigorously. "In brief, ONE is a symbol of the union of things: it introduces that generating virtue which is the cause of all combinations; and consequently ONE is a good principle."

"Ah, ah, ah!" said Queen Dolores, by and by. "Ah, I heartily admire a good principle. But what, alas, has now become of your concrete example?"

"It stands in readiness, madame: there is but ONE Jurgen."

"Oh, I assure you, I am not yet convinced of that. Still, the surprising audacity of your example will, I confess, help me to remember ONE, whether or not you prove to be really unique."

"Now, Two, or the dyad, the origin of contrasts—"

Jurgen went on penetratingly to demonstrate that Two was a symbol of diversity and of restlessness and of disorder, ending in collapse and separation; and was accordingly an evil principle. Thus was the life of every man made wretched by the struggle between his Two components, his soul and his body; and thus was the rapture of expectant parents considerably abated by the advent of TWINS.

THREE, or the triad, however, since everything was composed of three substances, contained the most sublime mysteries, which Jurgen duly communicated. We must remember, he pointed out, that Zeus carried a TRIPLE thunderbolt, and Poseidon a TRIDENT, whereas Adês was guarded by a dog with THREE heads: this in addition to the omnipotent brothers themselves being a TRIO.

Thus Jurgen continued to impart the Praxagorean significance of each digit separately; and by and by the Queen was declaring his flow of wisdom to be superhuman.

"Ah, but, madame, not even the wisdom of a king is without limit," says Jurgen modestly, "and no monarch may hope always to rise to every occasion. Kings, therefore, however properly anointed, may have their failings in strict uprightness, like any common man—"

"But you, Jurgen, are not a king. You are a dynasty."

"Dynasties also, madame, have by and by their tragic downfalls, as history over and over again attests. Meanwhile EIGHT, I repeat, is appropriately the number of the Beatitudes. And NINE, or the ennead, also, being the multiple of THREE, should be regarded as sacred—"

The Queen attended docilely to his demonstration of the peculiar properties of NINE. And when he had ended she confessed that beyond doubt NINE should be regarded as miraculous. But she repudiated his analogues as to the muses, the lives of a cat, and how many tailors made a man.

"Rather, I shall remember always," she declared, "that King Jurgen of Eubonia is a NINE days' wonder."

"Well, madame," said Jurgen, with a sigh, "now that we have reached NINE, I regret to say we have exhausted the digits."

"Oh, what a pity!" cried Queen Dolores. "Nevertheless, I will concede the only illustration I disputed; there is but ONE Jurgen; and certainly this Praxagorean system of mathematics is a fascinating study." And promptly she commenced to plan Jurgen's return with her into Philistia, so that she might perfect herself in the higher branches of mathematics. "For you must teach me calculus and geometry and all other sciences in which these digits are employed. We can arrange some compromise with the priests. That is always possible with the priests of Philistia, and indeed the priests of Sesphra can be made to help anybody in anything. And as for your Hamadryad, I will attend to her myself."

"But, no," says Jurgen, "I am ready enough in all conscience to compromise elsewhere; but to compound with the forces of Philistia is one thing I cannot do."

"Do you mean that, King Jurgen?" The Queen was astounded.

"I mean it, my dear, as I mean nothing else. You are in many ways an admirable people, and you are in all ways a formidable people. So I admire, I dread, I avoid, and at the very last pinch I defy. For you are not my people, and willy-nilly my gorge rises against your laws, as equally insane and abhorrent. Mind you, though, I assert nothing. You may be right in attributing wisdom to these laws; and certainly I cannot go so far as to say you are wrong: but still, at the same time—! That is the way I feel about it. So I, who compromise with everything else, can make no compromise with Philistia. No, my adored Dolores, it is not a virtue, rather it is an instinct with me, and I have no choice."

Even Dolores, who was Queen of all the Philistines, could perceive that this man spoke truthfully.

"I am sorry," says she, with real regret, "for you could be much run after in Philistia."

"Yes," said Jurgen, "as an instructor in mathematics."

"But, no, King Jurgen, not only in mathematics," said Dolores, reasonably. "There is poetry, for instance! For they tell me you are a poet, and a great many of my people take poetry quite seriously, I believe. Of course, I do not have much time for reading, myself. So you can be the Poet Laureate of Philistia, on any salary you like. And you can teach us all your ideas by writing beautiful poems about them. And you and I can be very happy together."

"Teach, teach! there speaks Philistia, and quite temptingly, too, through an adorable mouth, that would bribe me with praise and fine food and soft days forever. It is a thing that happens rather often, though. And I can but repeat that art is not a branch of pedagogy!"

"Really I am heartily sorry. For apart from mathematics, I like you, King Jurgen, just as a person."

"I, too, am sorry, Dolores. For I confess to a weakness for the women of Philistia."

"Certainly you have given me no cause to suspect you of any weakness in that quarter," observed Dolores, "in the long while you have been alone with me, and have talked so wisely and have reasoned so deeply. I am afraid that after tonight I shall find all other men more or less superficial. Heigho! and I shall probably weep my eyes out tomorrow when you are relegated to limbo. For that is what the priests will do with you, King Jurgen, on one plea or another, if you do not conform to the laws of Philistia."

"And that one compromise I cannot make! Ah, but even now I have a plan wherewith to escape your priests; and failing that, I possess a cantrap to fall back upon in my hour of direst need. My private affairs are thus not yet in a hopeless or even in a dejected condition. This fact now urges me to observe that Ten, or the decade, is the measure of all, since it contains all the numeric relations and harmonies—"

So they continued their study of mathematics until it was time for Jurgen to appear again before his judges. And in the morning Queen Dolores sent word to her priests that she was too sleepy to attend their council, but that the man was indisputably flesh and blood, amply deserved to be a king, and as a mathematician had not his peer.[1]

[1] *Note:* In a forgotten novel by the forgotten Victorian novelist Mrs. Henry Wood (who wrote *East Lynne*) there is an anticipation, suitable for the purer-minded, of Mr. Cabell's mathematical exposition. It goes like this:

> Edgar to Pleasaunce: "Now I kiss you three times on one cheek, and four times on your mouth. How many did that make altogether?" "Seven," whispered the girl, disengaging herself to breathe more freely. "That is arithmetic," said the youth triumphantly. "Dear me," said Pleasaunce, "I should not have thought it."

H. G. WELLS

# Peter Learns Arithmetic

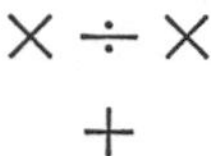

THEN MISS MILLS taught Peter to add and subtract and multiply and divide. She had once heard some lectures upon teaching arithmetic by graphic methods that had pleased her very much. They had seemed so clear. The lecturer had suggested that for a time easy sums might be shown in the concrete as well as in figures. You would first of all draw your operation or express it by wood blocks, and then you would present it in figures. You would draw an addition of 3 to 4, thus:

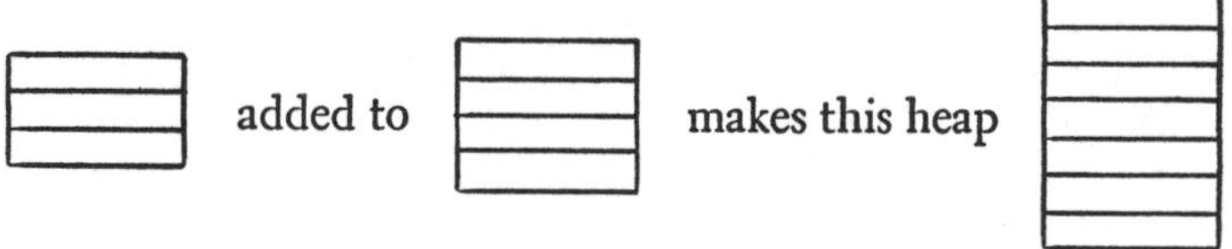


And then when your pupil had counted it and verified it you would write it down:

$$3 \quad + \quad 4 \quad = \quad 7$$

But Miss Mills, when she made her notes, had had no time to draw all the parallelograms; she had just put down one and a number over it in each case, and then her memory had muddled the idea. So she taught Joan and Peter thus: "See," she said, "I will make it perfectly plain to you. Perfectly plain. You take three—so," and she drew

| 3 |
|---|

"and then you take four—so," and she drew

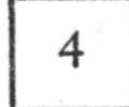

"and then you see three plus four makes seven—so:

$$\boxed{3} + \boxed{4} = \boxed{7}$$

"Do you see now how it *must* be so, Peter?"

Peter tried to feel that he did.

Peter quite agreed that it was nice to draw frames about the figures in this way. Afterward he tried a variation that looked like the face of old Chester Drawers:

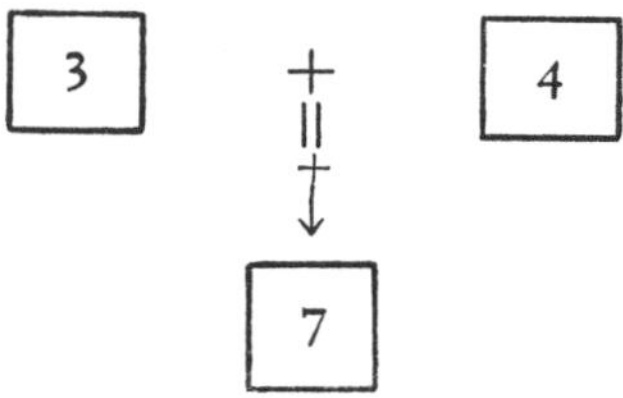

But for some reason Miss Mills would not see the beauty of that. Instead of laughing, she said, "Oh, no, that's *quite* wrong!" which seemed to Peter just selfishly insisting on her own way.

Peter was rather good at arithmetic, in spite of Miss Mills' instruction. He got sums right. It was held to be a gift. Joan was less fortunate. Like most people who have been badly taught, Miss Mills had one or two foggy places in her own arithmetical equipment. She was not clear about seven sevens and eight eights; she had a confused, irregular tendency to think that they might amount in either case to fifty-six, and also she had a trick of adding seven to nine as fifteen, although she always got from nine to seven correctly as sixteen. Every learner of arithmetic has a tendency to start little local flaws of this sort, standing sources of error, and every good, trained teacher looks out for them, knows how to test for them and set them right. Once they have been faced in a clear-headed way, such flaws can be cured in an hour or so. But few teachers in upper and middle-class schools in England, in those days, knew even the elements of their business; and it was the custom to let the baffling influence of such flaws develop into the persuasion that the pupil had not "the gift for mathematics." Very few women indeed of the English "educated" classes to this day can understand a fraction or do an ordinary multiplication sum. They think computation is a sort of fudging—in which some people are persistently lucky enough to guess right—"the gift for mathematics"—or impudent enough to carry their points. That was Miss Mills' secret and unformulated conviction, a conviction with which she was infecting a large proportion of the youngsters committed to her care. Joan became a mathematical gambler of the wildest description. But there was a guiding light in Peter's little head that made him grip at last upon the conviction that seven sevens make always forty-nine, and eight eights always sixty-four, and that when this haunting fifty-six flapped about in the sums it was because Miss Mills, grown-up teacher though she was, was wrong.

PLATO

# Socrates and the Slave

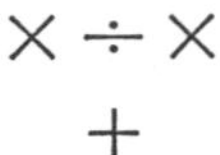

MENO: What do you mean by saying that we do not learn, and that what we call learning is only a process of recollection? Can you teach me how this is?

SOCRATES: I told you, Meno, just now that you were a rogue, and now you ask whether I can teach you, when I am saying that there is no teaching, but only recollection; and thus you imagine that you will involve me in a contradiction.

MENO: Indeed, Socrates, I protest that I had no such intention. I only asked the question from habit; but if you can prove to me that what you say is true, I wish that you would.

SOCRATES: It will be no easy matter, but I will try to please you to the utmost of my power. Suppose that you call one of your numerous attendants, that I may demonstrate on him.

MENO: Certainly. Come hither, boy.

SOCRATES: He is Greek, and speaks Greek, does he not?

MENO: Yes, indeed; he was born in the house.

SOCRATES: Attend now to the questions which I ask him, and observe whether he learns of me or only remembers.

MENO: I will.

SOCRATES: Tell me, boy, do you know that a figure like this is a square?

BOY: I do.

SOCRATES: And you know that a square figure has these four lines equal?

[FROM THE *Meno* of Plato, TRANSLATED BY BENJAMIN JOWETT, OXFORD UNIVERSITY PRESS, LONDON]

BOY: Certainly.

SOCRATES: And these lines which I have drawn through the middle of the square are also equal?

BOY: Yes.

SOCRATES: A square may be of any size?

BOY: Certainly.

SOCRATES: And if one side of the figure be of two feet, and the other side be of two feet, how much will the whole be? Let me explain: If in one direction the space was of two feet, and in the other direction of one foot, the whole would be of two feet taken once?

BOY: Yes.

SOCRATES: But since this side is also of two feet, there are twice two feet?

BOY: There are.

SOCRATES: Then the square is of twice two feet?

BOY: Yes.

SOCRATES: And how many are twice two feet? Count and tell me.

BOY: Four, Socrates.

SOCRATES: And might there not be another square twice as large as this, and having like this the lines equal?

BOY: Yes.

SOCRATES: And of how many feet will that be?

BOY: Of eight feet.

SOCRATES: And now try and tell me the length of the line which forms the side of that double square: this is two feet—what will that be?

BOY: Clearly, Socrates, it will be double.

SOCRATES: Do you observe, Meno, that I am not teaching the boy anything, but only asking him questions; and now he fancies that he knows how long a line is necessary in order to produce a figure of eight square feet, does he not?

MENO: Yes.

SOCRATES: And does he really know?

MENO: Certainly not.

SOCRATES: He only guesses that because the square is double, the line is double.

MENO: True.

SOCRATES: Observe him while he recalls the steps in regular order.

(*To the Boy.*) Tell me, boy, do you assert that a double space comes from a double line? Remember that I am not speaking of an oblong, but of a figure equal every way, and twice the size of this—that is to say of eight feet; and I want to know whether you still say that a double square comes from a double line?

BOY: Yes.

SOCRATES: But does not this line become doubled if we add another such line here?

BOY: Certainly.

SOCRATES: And four such lines will make a space containing eight feet?

BOY: Yes.

SOCRATES: Let us describe such a figure: Would you not say that this is the figure of eight feet?

BOY: Yes.

SOCRATES: And are there not these four divisions in the figure, each of which is equal to the figure of four feet?

BOY: True.

SOCRATES: And is not that four times four?

BOY: Certainly.

SOCRATES: And four times is not double?

BOY: No, indeed.

SOCRATES: But how much?

BOY: Four times as much.

SOCRATES: Therefore the double line, boy, has given a space, not twice, but four times as much.

BOY: True.

SOCRATES: Four times four are sixteen—are they not?

BOY: Yes.

SOCRATES: What line would give you a space of eight feet, as this gives one of sixteen feet;—do you see?

BOY: Yes.

SOCRATES: And the space of four feet is made from this half line?

BOY: Yes.

SOCRATES: Good; and is not a space of eight feet twice the size of this, and half the size of the other?

BOY: Certainly.

SOCRATES: Such a space, then, will be made out of a line greater than this one, and less than that one?

BOY: Yes; I think so.

SOCRATES: Very good; I like to hear you say what you think. And now tell me, is not this a line of two feet and that of four?

BOY: Yes.

SOCRATES: Then the line which forms the side of eight feet ought to be more than this line of two feet, and less than the other of four feet?

BOY: It ought.

SOCRATES: Try and see if you can tell me how much it will be.

BOY: Three feet.

SOCRATES: Then if we add a half to this line of two, that will be the line of three. Here are two and there is one; and on the other side, here are two also and there is one: and that makes the figure of which you speak?

BOY: Yes.

SOCRATES: But if there are three feet this way and three feet that way, the whole space will be three times three feet?

BOY: That is evident.

SOCRATES: And how much are three times three feet?

BOY: Nine.

SOCRATES: And how much is the double of four?

BOY: Eight.

SOCRATES: Then the figure of eight is not made out of a line of three?

BOY: No.

SOCRATES: But from what line?—tell me exactly; and if you would rather not reckon, try and show me the line.

BOY: Indeed, Socrates, I do not know.

SOCRATES: Do you see, Meno, what advances he has made in his power of recollection? He did not know at first, and he does not know now, what is the side of a figure of eight feet; but then he thought that he knew, and answered confidently as if he knew, and had no difficulty; now he has a difficulty, and neither knows nor fancies that he knows.

MENO: True.

SOCRATES: Is he not better off in knowing his ignorance?

MENO: I think that he is.

SOCRATES: If we have made him doubt, and given him the "torpedo's shock," have we done him any harm?

MENO: I think not.

SOCRATES: We have certainly, it would seem, assisted him in some degree to the discovery of the truth; and now he will wish to remedy his ignorance, but then he would have been ready to tell all the world again and again that the double space should have a double side.

MENO: True.

SOCRATES: But do you suppose that he would ever have inquired into or learned what he fancied that he knew, though he was really ignorant of it, until he had fallen into perplexity under the idea that he did not know, and had desired to know?

MENO: I think not, Socrates.

SOCRATES: Then he was the better for the torpedo's touch?

MENO: I think so.

SOCRATES: Mark now the farther development. I shall only ask him, and not teach him, and he shall share the inquiry with me: and do you watch and see if you find me telling or explaining anything to him, instead of eliciting his opinion. Tell me, boy, is not this a square of four feet which I have drawn?

BOY: Yes.

SOCRATES: And now I add another square equal to the former one?

BOY: Yes.

SOCRATES: And a third, which is equal to either of them?

BOY: Yes.

SOCRATES: Suppose that we fill up the vacant corner?

BOY: Very good.

SOCRATES: Here, then, there are four equal spaces?

BOY: Yes.

SOCRATES: And how many times larger is this space than this other?

BOY: Four times.

SOCRATES: But it ought to have been twice only, as you will remember.

BOY: True.

SOCRATES: And do not these lines, reaching from corner to corner, bisect each of these spaces?

BOY: Yes.

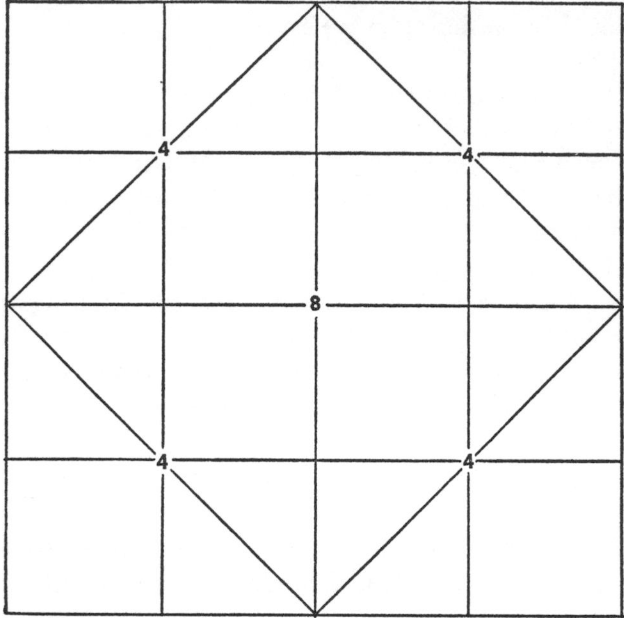


SOCRATES: And are there not here four equal lines which contain this space?

BOY: There are.

SOCRATES: Look and see how much this space is.

BOY: I do not understand.

SOCRATES: Has not each interior line cut off half of the four spaces?

BOY: Yes.

SOCRATES: And how many spaces are there in this section?

BOY: Four.

SOCRATES: And how many in this?

BOY: Two.

SOCRATES: And four is how many times two?

BOY: Twice.

SOCRATES: And this space is of how many feet?

BOY: Of eight feet.

SOCRATES: And from what line do you get this figure?

BOY: From this.

SOCRATES: That is, from the line which extends from corner to corner of the figure of four feet?

BOY: Yes.

SOCRATES: And that is the line which the learned call the diagonal. And if this is the proper name, then you, Meno's slave, are prepared to affirm that the double space is the square of the diagonal?

BOY: Certainly, Socrates.

SOCRATES: What do you say of him, Meno? Were not all these answers given out of his own head?

MENO: Yes, they were all his own.

SOCRATES: And yet, as we were just now saying, he did not know?

MENO: True.

SOCRATES: But still he had in him those notions of his—had he not?

MENO: Yes.

SOCRATES: Then he who does not know may still have true notions of that which he does not know?

MENO: He has.

SOCRATES: And at present these notions have just been stirred up in him, as in a dream; but if he were frequently asked the same questions, in different forms, he would know as well as any one at last?

MENO: I dare say.

SOCRATES: Without anyone teaching him he will recover his knowledge for himself, if he is only asked questions?

MENO: Yes.

SOCRATES: And this spontaneous recovery of knowledge in him is recollection?

MENO: True.

SOCRATES: And this knowledge which he now has must he not either have acquired or always possessed?

MENO: Yes.

SOCRATES: But if he always possessed this knowledge he would always have known; or if he has acquired the knowledge he could not have acquired it in this life, unless he has been taught geometry; for he may be made to do the same with all geometry and every other branch of knowledge. Now, has anyone ever taught him all this? You must know about him, if, as you say, he was born and bred in your house.

MENO: And I am certain that no one ever did teach him.

SOCRATES: And yet he has the knowledge?

MENO: The fact, Socrates, is undeniable.

SOCRATES: But if he did not acquire the knowledge in this life, then he must have had and learned it at some other time?

MENO: Clearly he must.

SOCRATES: Which must have been the time when he was not a man?

MENO: Yes.

SOCRATES: And if there have been always true thoughts in him, both at the time when he was and was not a man, which only need to be awakened into knowledge by putting questions to him, his soul must have always possessed this knowledge, for he always either was or was not a man?

MENO: Obviously.

SOCRATES: And if the truth of all things always existed in the soul, then the soul is immortal. Wherefore be of good cheer, and try to recollect what you do not know, or rather what you do not remember.

KAREL ČAPEK

# The Death of Archimedes

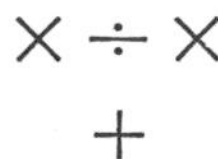

THE STORY OF ARCHIMEDES did not happen quite in the way that has been written; it is true that he was killed when the Romans conquered Syracuse, but it is not correct that a Roman soldier burst into his house to plunder it and that Archimedes, absorbed in drawing a geometrical figure, growled at him crossly: "Don't spoil my circles!" For one thing, Archimedes was not an absent-minded professor who did not know what was going on around him; on the contrary, he was by nature a thorough soldier who invented engines of war for the Syracusans for the defense of the city; for another thing, the Roman soldier was not a drunken plunderer but the educated and ambitious staff centurion Lucius, who knew to whom he had the honor of speaking and had not come to plunder but saluted on the threshold and said: "Greeting, Archimedes."

Archimedes raised his eyes from the wax tablet on which he was in fact drawing something and said, "What is it?"

"Archimedes," said Lucius, "we know that without your engines of war Syracuse wouldn't have held out for a month; as it is, we have had our hands full with it for two years. Don't imagine that we soldiers don't appreciate that. They're magnificent engines. My congratulations."

Archimedes waved his hand. "Please don't, they're really nothing. Just ordinary mechanisms for throwing projectiles—mere toys. From a scientific point of view they have no great importance."

"But from a military one they have," said Lucius. "Listen, Archimedes, I have come to ask you to work with us."

"With whom?"

"With us, the Romans. After all, you must know that Carthage is on the decline. What is the use of helping her? We shall soon have the Carthaginians on the run, you'll see. You'd better join us, all of you."

"Why?" growled Archimedes. "We Syracusans happen to be Greeks. Why should we join you?"

"Because you live in Sicily, and we need Sicily."

"And why do you need it?"

"Because we want to be masters of the Mediterranean."

"Aha," said Archimedes, and stared reflectively at his tablet. "And why do you want that?"

"Whoever is master of the Mediterranean," said Lucius, "is master of the world. That's clear enough."

"And must you be masters of the world?"

"Yes. The mission of Rome is to become master of the world. And I can tell you, that's what Rome is going to be."

"Possibly," said Archimedes and erased something on his tablet. "But I wouldn't advise it, Lucius. Listen, to be master of the world—some day that's going to give you an awful lot of defending to do. It's a pity, all the trouble you're going to have with it."

"That doesn't matter; we shall be a great Empire."

"A great empire," murmured Archimedes. "If I draw a small circle or a large circle, it's still only a circle. There are still frontiers—you will never be without frontiers, Lucius. Do you think that a large circle is more perfect than a small circle? Do you think you are a greater geometrician if you draw a larger circle?"

"You Greeks are always juggling with arguments," objected the centurion. "We have another way of proving that we are right."

"How?"

"By action. For instance, we have conquered your Syracuse. Ergo, Syracuse belongs to us. Is that a clear proof?"

"Yes," said Archimedes and scratched his head with his stylo. "Yes, you have conquered Syracuse; only it is not and never will be the same Syracuse as it was before. Why, it used to be a great and famous city; now it will never be great again. Poor Syracuse!"

"But Rome will be great. Rome has got to be stronger than anyone else in the whole world."

"Why?"

"To keep up her position. The stronger we are, the more enemies we have. That is why we must be the strongest."

"As for strength," murmured Archimedes, "I'm a bit of a physicist, Lucius, and I'll tell you something. Force absorbs itself."

"What does that mean?"

"It's just a law, Lucius. Force which is active absorbs itself. The stronger you are, the more of your strength you use up that way; and one day a time will come—"

"What were you going to say?"

"Oh, nothing. I'm not a prophet, Lucius; I'm only a physicist. Force absorbs itself. I know no more than that."

"Listen, Archimedes, wouldn't you like to work with us? You have no idea what tremendous possibilities would open out for you in Rome. You would build the strongest war machines in the world—"

"Forgive me, Lucius; I'm an old man and I should like to work out one or two of my ideas. As you see, I am just drawing something here."

"Archimedes, aren't you attracted by the idea of winning world mastery with us? —Why don't you answer?"

"I beg your pardon," grunted Archimedes, bending over his tablets. "What did you say?"

"That a man like you might win world mastery."

"Hm, world mastery," said Archimedes in a bored tone. "You mustn't be offended, but I've something more important here. Something more lasting, you know. Something which will really endure."

"What's that?"

"Mind! Don't spoil my circles! It's the method of calculating the area of a segment of a circle."

* * *

Later it was reported that the learned Archimedes had lost his life through an accident.

# II
# IMAGINARIES

ARTHUR PORGES

# The Devil and Simon Flagg

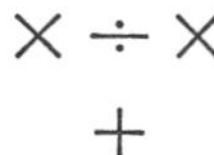

After several months of the most arduous research, involving the study of countless faded manuscripts, Simon Flagg succeeded in summoning the devil. As a competent medievalist, his wife had proved invaluable. A mere mathematician himself, he was hardly equipped to decipher Latin holographs, particularly when complicated by rare terms from tenth-century demonology, so it was fortunate that she had a flair for such documents.

The preliminary skirmishing over, Simon and the devil settled down to bargain in earnest. The devil was sulky, for Simon had scornfully declined several of his most dependable gambits, easily spotting the deadly barb concealed in each tempting bait.

"Suppose you listen to a proposition from me for a change," Simon suggested finally. "At least, it's a straightforward one."

The devil irritably twirled his tail-tip with one hand, much as a man might toy with his key chain. Obviously, he felt injured.

"All right," he agreed, in a grumpy voice. "It can't do any harm. Let's hear your proposal."

"I will pose a certain question," Simon began, and the devil brightened, "to be answered within twenty-four hours. If you cannot do so, you must pay me $100,000. That's a modest request compared to most you get. No billions, no Helen of Troy on a tiger skin. Naturally there must be no reprisals of any kind if I win."

"Indeed!" the devil snorted. "And what are your stakes?"

"If I lose, I will be your slave for any short period. No torment, no loss of soul—not for a mere $100,000. Neither will I harm relatives or friends. Although," he amended thoughtfully, "there are exceptions."

The devil scowled, pulling his forked tail petulantly. Finally, a savage tug having brought a grimace of pain, he desisted.

"Sorry," he said flatly. "I deal only in souls. There is no shortage of slaves. The amount of free, wholehearted service I receive from humans would amaze you. However, here's what I'll do. If I can't answer your question in the given time, you will receive not a paltry $100,000, but any sum within reason. In addition, I offer health and happiness as long as you live. If I do answer it—well, you know the consequences. That's the very best I can offer." He pulled a lighted cigar from the air and puffed in watchful silence.

Simon stared without seeing. Little moist patches sprang out upon his forehead. Deep in his heart he had known what the devil's only terms would be. Then his jaw muscles knotted. He would stake his soul that nobody—man, beast, or devil—could answer *this* question in twenty-four hours.

"Include my wife in that health and happiness provision, and it's a deal," he said. "Let's get on with it."

The devil nodded. He removed the cigar stub from his mouth, eyed it distastefully, and touched it with a taloned forefinger. Instantly it became a large pink mint, which he sucked with noisy relish.

"About your question," he said, "it must have an answer, or our contract becomes void. In the Middle Ages, people were fond of proposing riddles. A few came to me with paradoxes, such as that one about a village with one barber who shaves all those, and only those, who don't shave themselves. 'Who shaves the barber?' they asked. Now, as Russell has noted, the 'all' makes such a question meaningless and so unanswerable."

"My question is just that—not a paradox," Simon assured him.

"Very well. I'll answer it. What are you smirking about?"

"Nothing," Simon replied, composing his face.

"You have very good nerves," the devil said, grimly approving, as he pulled a parchment from the air. "If I had chosen to appear as a certain monster which combines the best features of your gorilla with those of the Venusian Greater Kleep, an animal—I suppose one could call it that—of unique eye appeal, I wonder if your aplomb—"

"You needn't make any tests," Simon said hastily. He took the proffered contract and, satisfied that all was in order, opened his pocketknife.

"Just a moment," the devil protested. "Let me sterilize that; you might get infected." He held the blade to his lips, blew gently, and the steel glowed cherry red. "There you are. Now a touch of the point to some—ah—ink, and we're all set. Second line from the bottom, please; the last one's mine."

Simon hesitated, staring at the moist red tip.

"Sign," urged the devil, and, squaring his shoulders, Simon did so.

When his own signature had been added with a flourish, the devil rubbed his palms together, gave Simon a frankly proprietary glance, and said jovially, "Let's have the question. As soon as I answer it, we'll hurry off. I've just time for another client tonight."

"All right," said Simon. He took a deep breath. "My question is this: Is Fermat's Last Theorem correct?"

The devil gulped. For the first time his air of assurance weakened.

"Whose last what?" he asked in a hollow voice.

"Fermat's Last Theorem. It's a mathematical proposition which Fermat, a seventeenth-century French mathematician, claimed to have proved. However, his proof was never written down, and to this day nobody knows if the theorem is true or false." His lips twitched briefly as he saw the devil's expression. "Well, there you are—go to it!"

"Mathematics!" the devil exclaimed, horrified. "Do you think I've had time to waste learning such stuff? I've studied the Trivium and Quadrivium, but as for algebra—say," he added resentfully, "what kind of a question is that to ask me?"

Simon's face was strangely wooden, but his eyes shone. "You'd rather run 75,000 miles and bring back some object the size of Boulder Dam, I suppose!" he jeered. "Time and space are easy for you, aren't they? Well, sorry. I prefer this. It's a simple matter," he added, in a bland voice. "Just a question of positive integers."

"What's a positive integer?" the devil flared. "Or an integer, for that matter?"

"To put it more formally," Simon said, ignoring the devil's question, "Fermat's Theorem states that there are no non-trivial, rational solutions of the equation $X^n + Y^n = Z^n$, for $n$ a positive integer greater than two."

"What's the meaning of—"

"You supply the answers, remember."

"And who's to judge—you?"

"No," Simon replied sweetly. "I doubt if I'm qualified, even after studying the problem for years. If you come up with a solution, we'll submit it to any good mathematical journal, and their referee will decide. And you can't back out—the problem obviously is soluble: either the theorem is true, or it is false. No nonsense about multivalued logic, mind. Merely determine which, and *prove* it in twenty-four hours. After all, a man—excuse me—demon, of your intelligence and vast experience surely can pick up a little math in that time."

"I remember now what a bad time I had with Euclid when I studied at Cambridge," the devil said sadly. "My proofs were always wrong, and yet it was all obvious anyway. You could see just by the diagrams." He set his jaw. "But I can do it. I've done harder things before. Once I went to a distant star and brought back a quart of neutronium in just sixteen—"

"I know," Simon broke in. "You're very good at such tricks."

"Trick, nothing!" was the angry retort. "It's a technique so difficult —but never mind, I'm off to the library. By this time tomorrow—"

"No," Simon corrected him. "We signed half an hour ago. Be back in exactly twenty-three point five hours! Don't let me rush you," he added ironically, as the devil gave the clock a startled glance. "Have a drink and meet my wife before you go."

"I never drink on duty. Nor have I time to make the acquaintance of your wife . . . now." He vanished.

The moment he left, Simon's wife entered.

"Listening at the door again?" Simon chided her, without resentment.

"Naturally," she said in her throaty voice. "And darling—I want to know—that question—is it really difficult? Because if it's not—Simon, I'm so worried."

"It's difficult, all right." Simon was almost jaunty. "But most people don't realize that at first. You see," he went on, falling automatically into his stance for Senior Math II, "anybody can find two whole numbers whose squares add up to a square. For example, $3^2 + 4^2 = 5^2$; that is, $9 + 16 = 25$. See?"

"Uh huh." She adjusted his tie.

"But when you try to find two cubes that add up to a cube, or higher powers that work similarly, there don't seem to be any. Yet," he concluded dramatically, "nobody has been able to prove that no such numbers exist. Understand now?"

"Of course." Simon's wife always understood mathematical statements, however abstruse. Otherwise, the explanation was repeated until she did, which left little time for other activities.

"I'll make us some coffee," she said, and escaped.

Four hours later as they sat together listening to Brahms' Third, the devil reappeared.

"I've already learned the fundamentals of algebra, trigonometry, and plane geometry!" he announced triumphantly.

"Quick work," Simon complimented him. "I'm sure you'll have no trouble at all with spherical, analytic, projective, descriptive, and non-Euclidean geometries."

The devil winced. "Are there so many?" he inquired in a small voice.

"Oh, those are only a few." Simon had the cheerful air suited to a bearer of welcome tidings. "You'll like non-Euclidean," he said mendaciously. "There you don't have to worry about diagrams—they don't tell a thing! And since you hated Euclid anyway—"

With a groan the devil faded out like an old movie. Simon's wife giggled.

"Darling," she sang, "I'm beginning to think you've got him over a barrel."

"Sh," said Simon. "The last movement. Glorious!"

Six hours later, there was a smoky flash, and the devil was back. Simon noted the growing bags under his eyes. He suppressed a grin.

"I've learned all those geometries," the devil said with grim satisfaction. "It's coming easier now. I'm about ready for your little puzzle."

Simon shook his head. "You're trying to go too fast. Apparently you've overlooked such basic techniques as calculus, differential equations, and finite differences. Then there's—"

"Will I need all those?" the devil moaned. He sat down and knuckled his puffy eyelids, smothering a yawn.

"I couldn't say," Simon replied, his voice expressionless. "But people have tried practically every kind of math there is on that 'little puzzle,' and it's still unsolved. Now, I suggest—" But the devil was in no mood for advice from Simon. This time he even made a sloppy disappearance while sitting down.

"I think he's tired," Mrs. Flagg said. "Poor devil." There was no discernible sympathy in her tones.

"So am I," said Simon. "Let's get to bed. He won't be back until tomorrow, I imagine."

"Maybe not," she agreed, adding demurely, "but I'll wear the black lace—just in case."

It was the following afternoon. Bach seemed appropriate somehow, so they had Landowska on.

"Ten more minutes," Simon said. "If he's not back with a solution by then, we've won. I'll give him credit; he could get a Ph.D. out of my school in one day—with honors! However—"

There was a hiss. Rosy clouds mushroomed sulphurously. The devil stood before them, steaming noisomely on the rug. His shoulders sagged; his eyes were bloodshot; and a taloned paw, still clutching a sheaf of papers, shook violently from fatigue or nerves.

Silently, with a kind of seething dignity, he flung the papers to the floor, where he trampled them viciously with his cloven hoofs. Gradually then, his tense figure relaxed, and a wry smile twisted his mouth.

"You win, Simon," he said, almost in a whisper, eying him with ungrudging respect. "Not even I can learn enough mathematics in such a short time for so difficult a problem. The more I got into it, the worse it became. Non-unique factoring, ideals— Baal! Do you know," he confided, "not even the best mathematicians on other planets—all far ahead of yours—have solved it? Why, there's a chap on Saturn—he looks something like a mushroom on stilts—who solves partial differential equations mentally; and even he's given up." The devil sighed. "Farewell." He dislimned with a kind of weary precision.

Simon kissed his wife—hard. A long while later she stirred in his arms.

"Darling," she pouted, peering into his abstracted face, "what's wrong now?"

"Nothing—except I'd like to see his work; to know how close he came. I've wrestled with that problem for—" He broke off amazed as the devil flashed back. Satan seemed oddly embarrassed.

"I forgot," he mumbled. "I need to—ah!" He stooped for the scattered papers, gathering and smoothing them tenderly. "It certainly gets you," he said, avoiding Simon's gaze. "Impossible to stop just now. Why, if I could only prove one simple little lemma—" He saw the blazing interest in Simon, and dropped his apologetic air. "Say," he grunted, "you've worked on this, I'm sure. Did you try continued fractions? Fermat must have used them, and—move over a minute, please—" This last to Mrs. Flagg. He sat down beside Simon, tucked his tail under, and pointed to a jungle of symbols.

Mrs. Flagg sighed. Suddenly the devil seemed a familiar figure, little different from old Professor Atkins, her husband's colleague at the university. Any time two mathematicians got together on a tantalizing problem . . . Resignedly she left the room, coffeepot in hand. There was certainly a long session in sight. She knew. After all, she was a professor's wife.

ROBERT A. HEINLEIN

# —And He Built a Crooked House

× ÷ ×
+

AMERICANS ARE CONSIDERED crazy anywhere in the world.

They will usually concede a basis for the accusation but point to California as the focus of the infection. Californians stoutly maintain that their bad reputation is derived solely from the acts of the inhabitants of Los Angeles County. Angelenos will, when pressed, admit the charge but explain hastily, "It's Hollywood. It's not our fault —we didn't ask for it; Hollywood just grew."

The people in Hollywood don't care; they glory in it. If you are interested, they will drive you up Laurel Canyon "—where we keep the violent cases." The Canyonites—the brown-legged women, the trunks-clad men constantly busy building and rebuilding their slap-happy unfinished houses—regard with faint contempt the dull creatures who live down in the flats, and treasure in their hearts the secret knowledge that they, and only they, know how to live.

Lookout Mountain Avenue is the name of a side canyon which twists up from Laurel Canyon. The other Canyonites don't like to have it mentioned; after all, one must draw the line somewhere!

High up on Lookout Mountain at number 8775, across the street from the Hermit—the original Hermit of Hollywood—lived Quintus Teal, graduate architect.

Even the architecture of southern California is different. Hot dogs are sold from a structure built like and designated "The Pup." Ice-cream cones come from a giant stucco ice-cream cone, and neon proclaims "Get the Chili Bowl Habit!" from the roofs of buildings which are indisputably chili bowls. Gasoline, oil, and free road maps are dispensed beneath the wings of tri-motored transport planes, while the certified rest rooms, inspected hourly for your comfort, are located in the cabin of the plane itself. These things may surprise, or amuse, the tourist, but the local residents, who walk bareheaded in the famous California noonday sun, take them as a matter of course.

Quintus Teal regarded the efforts of his colleagues in architecture as fainthearted, fumbling, and timid.

"What is a house?" Teal demanded of his friend, Homer Bailey.

"Well—" Bailey admitted cautiously—"speaking in broad terms, I've always regarded a house as a gadget to keep off the rain."

"Nuts! You're as bad as the rest of them."

"I didn't say the definition was complete—"

"Complete! It isn't even in the right direction. From that point of view we might just as well be squatting in caves. But I don't blame you," Teal went on magnanimously, "you're no worse than the lugs you find practicing architecture. Even the Moderns—all they've done is to abandon the Wedding Cake School in favor of the Service Station School, chucked away the gingerbread and slapped on some chromium, but at heart they are as conservative and traditional as a county courthouse. Neutra! Schindler! What have those bums got? What's Frank Lloyd Wright got that I haven't got?"

"Commissions," his friend answered succinctly.

"Huh? Wha' d'ju say?" Teal stumbled slightly in his flow of words, did a slight double take, and recovered himself. "Commissions. Correct. And why? Because I don't think of a house as an upholstered cave; I think of it as a machine for living, a vital process, a live dynamic thing, changing with the mood of the dweller—not a dead, static, oversized coffin. Why should we be held down by the frozen concepts of our ancestors? Any fool with a little smattering of descriptive geometry can design a house in the ordinary way. Is the static geometry of Euclid the only mathematics? Are we to completely disregard the Picard-Vessiot theory? How about modular systems? To say nothing of the rich suggestions of stereochemistry. Isn't there a place in architecture for transformation, for homomorphology, for actional structures?"

"Blessed if I know," answered Bailey. "You might just as well be talking about the fourth dimension for all it means to me."

"And why not? Why should we limit ourselves to the— Say!" He interrupted himself and stared into distances. "Homer, I think you've really got something. After all, why not? Think of the infinite richness of articulation and relationship in four dimensions. What a house, what a house—" He stood quite still, his pale bulging eyes blinking thoughtfully.

Bailey reached up and shook his arm. "Snap out of it. What the hell are you talking about, four dimensions? Time is the fourth dimension; you can't drive nails into *that*."

Teal shrugged him off. "Sure. Sure. Time is *a* fourth dimension, but I'm thinking about a fourth spatial dimension, like length, breadth and thickness. For economy of materials and convenience of arrangement you couldn't beat it. To say nothing of the saving of ground space—you could put an eight-room house on the land now occupied by a one-room house. Like a tesseract—"

"What's a tesseract?"

"Didn't you go to school? A tesseract is a hypercube, a square figure with four dimensions to it, like a cube has three, and a square has two. Here, I'll show you." Teal dashed out into the kitchen of his apartment and returned with a box of toothpicks which he spilled on the table between them, brushing glasses and a nearly empty Holland gin bottle carelessly aside. "I'll need some plasticine. I had some around here last week." He burrowed into a drawer of the littered desk which crowded one corner of his dining room and emerged with a lump of oily sculptor's clay. "Here's some."

"What are you going to do?"

"I'll show you." Teal rapidly pinched off small masses of the clay and rolled them into pea-sized balls. He stuck toothpicks into four of these and hooked them together into a square. "There! That's a square."

"Obviously."

"Another one like it, four more toothpicks, and we make a cube." The toothpicks were now arranged in the framework of a square box, a cube, with the pellets of clay holding the corners together. "Now we make another cube just like the first one, and the two of them will be two sides of the tesseract."

Bailey started to help him roll the little balls of clay for the second cube, but became diverted by the sensuous feel of the docile clay and started working and shaping it with his fingers.

"Look," he said, holding up his effort, a tiny figurine, "Gypsy Rose Lee."

"Looks more like Gargantua; she ought to sue you. Now pay attention. You open up one corner of the first cube, interlock the second cube at one corner, and then close the corner. Then take eight more toothpicks and join the bottom of the first cube to the bottom of the second, on a slant, and the top of the first to the top of the second, the same way." This he did rapidly, while he talked.

"What's that supposed to be?" Bailey demanded suspiciously.

"That's a tesseract, eight cubes forming the sides of a hypercube in four dimensions."

"It looks more like a cat's cradle to me. You've only got two cubes there anyhow. Where are the other six?"

"Use your imagination, man. Consider the top of the first cube in relation to the top of the second; that's cube number three. Then the two bottom squares, then the front faces of each cube, the back faces, the right hand, the left hand—eight cubes." He pointed them out.

"Yeah, I see 'em. But they still aren't cubes; they're whatchamucallems—prisms. They are not square, they slant."

"That's just the way you look at it, in perspective. If you drew a picture of a cube on a piece of paper, the side squares would be slaunchwise, wouldn't they? That's perspective. When you look at a four-dimensional figure in three dimensions, naturally it looks crooked. But those are all cubes just the same."

"Maybe they are to you, brother, but they still look crooked to me."

Teal ignored the objections and went on. "Now consider this as the framework of an eight-room house; there's one room on the ground flour—that's for service, utilities, and garage. There are six rooms opening off it on the next floor, living room, dining room, bath, bedrooms, and so forth. And up at the top, completely enclosed and with windows on four sides, is your study. There! How do you like it?"

"Seems to me you have the bathtub hanging out of the living-room ceiling. Those rooms are interlaced like an octopus."

"Only in perspective, only in perspective. Here, I'll do it another way so you can see it." This time Teal made a cube of toothpicks, then made a second of halves of toothpicks, and set it exactly in the center of the first by attaching the corners of the small cube to the large cube by short lengths of toothpick. "Now—the big cube is your ground floor, the little cube inside is your study on the top floor. The six cubes joining them are the living rooms. See?"

Bailey studied the figure, then shook his head. "I still don't see but two cubes, a big one and a little one. Those other six things, they look like pyramids this time instead of prisms, but they still aren't cubes."

"Certainly, certainly, you are seeing them in different perspective. Can't you see that?"

"Well, maybe. But that room on the inside, there. It's completely surrounded by the thingamujigs. I thought you said it had windows on four sides."

"It has—it just looks like it was surrounded. That's the grand feature about a tesseract house, complete outside exposure for every room, yet every wall serves two rooms and an eight-room house requires only a one-room foundation. It's revolutionary."

"That's putting it mildly. You're crazy, Bud; you can't build a house like that. That inside room is on the inside, and there she stays."

Teal looked at his friend in controlled exasperation. "It's guys like you that keep architecture in its infancy. How many square sides has a cube?"

"Six."

"How many of them are inside?"

"Why, none of 'em. They're all on the outside."

"All right. Now listen—a tesseract has eight cubical sides, *all on the outside*. Now watch me. I'm going to open up this tesseract like you can open up a cubical pasteboard box, until it's flat. That way you'll be able to see all eight of the cubes." Working very rapidly he constructed four cubes, piling one on top of the other in an unsteady tower. He then built out four more cubes from the four exposed faces of the second cube in the pile. The structure swayed a little under the loose coupling of the clay pellets, but it stood, eight cubes in an inverted cross, a double cross, as the four additional cubes stuck out in four directions. "Do you see it now? It rests on the ground floor room, the next six cubes are the living rooms, and there is your study, up at the top."

Bailey regarded it with more approval than he had the other figures. "At least I can understand it. You say that is a tesseract, too?"

"That is a tesseract unfolded in three dimensions. To put it back together you tuck the top cube onto the bottom cube, fold those side cubes in till they meet the top cube and there you are. You do all this folding through a fourth dimension of course; you don't distort any of the cubes, or fold them into each other."

Bailey studied the wobbly framework further. "Look here," he said at last, "why don't you forget about folding this thing up through a fourth dimension—you can't anyway—and build a house like this?"

"What do you mean, I can't? It's a simple mathematical problem—"

"Take it easy, son. It may be simple in mathematics, but you could never get your plans approved for construction. There isn't any fourth dimension; forget it. But this kind of a house—it might have some advantages."

Checked, Teal studied the model. "Hm-m-m—Maybe you got something. We could have the same number of rooms, and we'd save the same amount of ground space. Yes, and we would set that middle cross-shaped floor northeast, southwest, and so forth, so that every room would get sunlight all day long. That central axis lends itself nicely to central heating. We'll put the dining room on the northeast and the kitchen on the southeast, with big view windows in every room. O.K., Homer, I'll do it! Where do you want it built?"

"Wait a minute! Wait a minute! I didn't say you were going to build it for me—"

"Of course I am. Who else? Your wife wants a new house; this is it."

"But Mrs. Bailey wants a Georgian house—"

"Just an idea she has. Women don't know what they want—"

"Mrs. Bailey does."

"Just some idea an out-of-date architect has put in her head. She drives a 1941 car, doesn't she? She wears the very latest styles—why should she live in an eighteenth-century house? This house will be even later than a 1941 model; it's years in the future. She'll be the talk of the town."

"Well—I'll have to talk to her."

"Nothing of the sort. We'll surprise her with it. Have another drink."

"Anyhow, we can't do anything about it now. Mrs. Bailey and I are driving up to Bakersfield tomorrow. The company's bringing in a couple of wells tomorrow."

"Nonsense. That's just the opportunity we want. It will be a surprise for her when you get back. You can just write me a check right now, and your worries are over."

"I oughtn't to do anything like this without consulting her. She won't like it."

"Say, who wears the pants in your family anyhow?"

The check was signed about halfway down the second bottle.

Things are done fast in southern California. Ordinary houses there are usually built in a month's time. Under Teal's impassioned heckling the tesseract house climbed dizzily skyward in days rather than weeks, and its cross-shaped second story came jutting out at the four corners of the world. He had some trouble at first with the inspectors over these four projecting rooms but by using strong girders and folding money he had been able to convince them of the soundness of his engineering.

By arrangement, Teal drove up in front of the Bailey residence the morning after their return to town. He improvised on his two-tone horn. Bailey stuck his head out the front door. "Why don't you use the bell?"

"Too slow," answered Teal cheerfully. "I'm a man of action. Is Mrs. Bailey ready? Ah, there you are, Mrs. Bailey! Welcome home, welcome home. Jump in, we've got a surprise for you!"

"You know Teal, my dear," Bailey put in uncomfortably.

Mrs. Bailey sniffed. "I know him. We'll go in our own car, Homer."

"Certainly, my dear."

"Good idea," Teal agreed; " 'sgot more power than mine; we'll get there faster. I'll drive, I know the way." He took the keys from Bailey, slid into the driver's seat, and had the engine started before Mrs. Bailey could rally her forces.

"Never have to worry about my driving," he assured Mrs. Bailey, turning his head as he did so, while he shot the powerful car down the avenue and swung onto Sunset Boulevard; "it's a matter of power and control, a dynamic process, just my meat—I've never had a serious accident."

"You won't have but one," she said bitingly. "Will you *please* keep your eyes on the traffic?"

He attempted to explain to her that a traffic situation was a matter, not of eyesight, but intuitive integration of courses, speeds, and probabilities, but Bailey cut him short. "Where is the house, Quintus?"

"House?" asked Mrs. Bailey suspiciously. "What's this about a house, Homer? Have you been up to something without telling me?"

Teal cut in with his best diplomatic manner. "It certainly is a house, Mrs. Bailey. And what a house! It's a surprise for you from a devoted husband. Just wait till you see it—"

"I shall," she agreed grimly. "What style is it?"

"This house sets a new style. It's later than television, newer than next week. It must be seen to be appreciated. By the way," he went on rapidly, heading off any retort, "did you folks feel the earthquake last night?"

"Earthquake? What earthquake? Homer, was there an earthquake?"

"Just a little one," Teal continued, "about two A.M. If I hadn't been awake, I wouldn't have noticed it."

Mrs. Bailey shuddered. "Oh, this awful country! Do you hear that, Homer? We might have been killed in our beds and never have known it. Why did I ever let you persuade me to leave Iowa?"

"But, my dear," he protested hopelessly, "you wanted to come out to California; you didn't like Des Moines."

"We needn't go into that," she said firmly. "You are a man; you should anticipate such things. Earthquakes!"

"That's one thing you needn't fear in your new home, Mrs. Bailey," Teal told her. "It's absolutely earthquake-proof; every part is in perfect dynamic balance with every other part."

"Well, I hope so. Where is this house?"

"Just around this bend. There's the sign now." A large arrow sign, of the sort favored by real-estate promoters, proclaimed in letters that were large and bright even for southern California:

THE HOUSE OF THE FUTURE!!!

COLOSSAL—AMAZING—REVOLUTIONARY

SEE HOW YOUR GRANDCHILDREN WILL LIVE!

Q. Teal, Architect

"Of course that will be taken down," he added hastily, noting her expression, "as soon as you take possession." He slued around the corner and brought the car to a squealing halt in front of the House of the Future. "*Voilà!*" He watched their faces for response.

Bailey stared unbelievingly, Mrs. Bailey in open dislike. They saw a simple cubical mass, possessing doors and windows, but no other architectural features, save that it was decorated in intricate mathematical designs. "Teal," Bailey asked slowly, "what have you been up to?"

Teal turned from their faces to the house. Gone was the crazy tower with its jutting second-story rooms. No trace remained of the seven rooms above ground floor level. Nothing remained but the single room that rested on the foundations. "Great jumping cats!" he yelled, "I've been robbed!"

He broke into a run.

But it did him no good. Front or back, the story was the same: the other seven rooms had disappeared, vanished completely. Bailey caught up with him and took his arm.

"Explain yourself. What is this about being robbed? How come you built anything like this—it's not according to agreement."

"But I didn't. I built just what we had planned to build, an eight-room house in the form of a developed tesseract. I've been sabotaged; that's what it is! Jealousy! The other architects in town didn't dare let me finish this job; they knew they'd be washed up if I did."

"When were you last here?"

"Yesterday afternoon."

"Everything all right then?"

"Yes. The gardeners were just finishing up."

Bailey glanced around at the faultlessly manicured landscaping. "I don't see how seven rooms could have been dismantled and carted away from here in a single night without wrecking this garden."

Teal looked around, too. "It doesn't look it. I don't understand it."

Mrs. Bailey joined them. "Well? Well? Am I to be left to amuse myself? We might as well look it over as long as we are here, though I'm warning you, Homer, I'm not going to like it."

"We might as well," agreed Teal, and drew a key from his pocket with which he let them in the front door. "We may pick up some clues."

The entrance hall was in perfect order, the sliding screens that separated it from the garage space were back, permitting them to see the entire compartment. "This looks all right," observed Bailey. "Let's go up on the roof and try to figure out what happened. Where's the staircase? Have they stolen that, too?"

"Oh, no," Teal denied, "look—" He pressed a button below the light switch; a panel in the ceiling fell away and a light, graceful flight of stairs swung noiselessly down. Its strength members were the frosty silver of duralumin, its treads and risers transparent plastic. Teal wriggled like a boy who has successfully performed a card trick, while Mrs. Bailey thawed perceptibly.

It was beautiful.

"Pretty slick," Bailey admitted. "Howsomever it doesn't seem to go any place—"

"Oh, that—" Teal followed his gaze. "The cover lifts up as you approach the top. Open stair wells are anachronisms. Come on." As predicted, the lid of the staircase got out of their way as they climbed the flight and permitted them to debouch at the top, but not, as they had expected, on the roof of the single room. They found themselves standing in the middle one of the five rooms which constituted the second floor of the original structure.

For the first time on record Teal had nothing to say. Bailey echoed him, chewing on his cigar. Everything was in perfect order. Before them, through open doorway and translucent partition lay the kitchen, a chef's dream of up-to-the-minute domestic engineering, Monel metal, continuous counter space, concealed lighting, functional arrangement. On the left the formal, yet gracious and hospitable, dining room awaited guests, its furniture in parade-ground alignment.

Teal knew before he turned his head that the drawing room and lounge would be found in equally substantial and impossible existence.

"Well, I must admit this *is* charming," Mrs. Bailey approved, "and the kitchen is just *too* quaint for words—though I would never have guessed from the exterior that this house had so much room upstairs. Of course *some* changes will have to be made. That secretary now—if we moved it over *here* and put the settle over *there*—"

"Stow it, Matilda," Bailey cut in brusquely. "What d'yuh make of it, Teal?"

"Why, Homer Bailey! The very id—"

"Stow it, I said. Well, Teal?"

The architect shuffled his rambling body. "I'm afraid to say. Let's go on up."

"How?"

"Like this." He touched another button; a mate, in deeper colors, to the fairy bridge that had let them up from below offered them access to the next floor. They climbed it, Mrs. Bailey expostulating in the rear, and found themselves in the master bedroom. Its shades were drawn, as had been those on the level below, but the mellow lighting came on automatically. Teal at once activated the switch which controlled still another flight of stairs, and they hurried up into the top floor study.

"Look, Teal," suggested Bailey when he had caught his breath, "can we get to the roof above this room? Then we could look around."

"Sure, it's an observatory platform." They climbed a fourth flight of stairs, but when the cover at the top lifted to let them reach the level above, they found themselves, not on the roof, but *standing in the ground floor room where they had entered the house.*

Mr. Bailey turned a sickly gray. "Angels in heaven," he cried, "this place is haunted. We're getting out of here." Grabbing his wife he threw open the front door and plunged out.

Teal was too much preoccupied to bother with their departure. There was an answer to all this, an answer that he did not believe. But he was forced to break off considering it because of hoarse shouts from somewhere above him. He lowered the staircase and rushed upstairs. Bailey was in the central room leaning over Mrs. Bailey, who had fainted. Teal took in the situation, went to the bar built into the lounge, and poured three fingers of brandy, which he returned with and handed to Bailey. "Here—this'll fix her up."

Bailey drank it.

"That was for Mrs. Bailey," said Teal.

"Don't quibble," snapped Bailey. "Get her another." Teal took the precaution of taking one himself before returning with a dose earmarked for his client's wife. He found her just opening her eyes.

"Here, Mrs. Bailey," he soothed, "this will make you feel better."

"I never touch spirits," she protested, and gulped it.

"Now tell me what happened," suggested Teal. "I thought you two had left."

"But we did—we walked out the front door and found ourselves up here, in the lounge."

"The hell you say! Hm-m-m—wait a minute." Teal went into the lounge. There he found that the big view window at the end of the room was open. He peered cautiously through it. He stared, not out at the California countryside, but into the ground floor room—or a reasonable facsimile thereof. He said nothing, but went back to the stair well which he had left open and looked down it. The ground floor room was still in place. Somehow, it managed to be in two different places at once, on different levels.

He came back into the central room and seated himself opposite Bailey in a deep, low chair, and sighted him past his upthrust bony knees. "Homer," he said impressively, "do you know what has happened?"

"No, I don't—but if I don't find out pretty soon, something is going to happen and pretty drastic, too!"

"Homer, this is a vindication of my theories. This house is a real tesseract."

"What's he talking about, Homer?"

"Wait, Matilda—now Teal, that's ridiculous. You've pulled some hanky-panky here and I won't have it—scaring Mrs. Bailey half to death, and making me nervous. All I want is to get out of here, with no more of your trapdoors and silly practical jokes."

"Speak for yourself, Homer," Mrs. Bailey interrupted, "I was *not* frightened; I was just took all over queer for a moment. It's my heart; all of my people are delicate and highstrung. Now about this tessy thing—explain yourself, Mr. Teal. Speak up."

He told her as well as he could in the face of numerous interruptions the theory back of the house. "Now as I see it, Mrs. Bailey," he concluded, "this house, while perfectly stable in three dimensions, was not stable in four dimensions. I had built a house in the shape of an unfolded tesseract; something happened to it, some jar or side thrust, and it collapsed into its normal shape—it folded up." He snapped his fingers suddenly. "I've got it! The earthquake!"

"Earthquake?"

"Yes, yes, the little shake we had last night. From a four-dimensional standpoint this house was like a plane balanced on edge. One little push and it fell over, collapsed along its natural joints into a stable four-dimensional figure."

"I thought you boasted about how safe this house was."

"It *is* safe—three-dimensionally."

"I don't call a house safe," commented Bailey edgily, "that collapses at the first little temblor."

"But look around you, man!" Teal protested. "Nothing has been disturbed, not a piece of glassware cracked. Rotation through a fourth dimension can't affect a three-dimensional figure any more than you can shake letters off a printed page. If you had been sleeping in here last night, you would never have awakened."

"That's just what I'm afraid of. Incidentally, has your great genius figured out any way for us to get out of this booby trap?"

"Huh? Oh, yes, you and Mrs. Bailey started to leave and landed back up here, didn't you? But I'm sure there is no real difficulty—we came in, we can go out. I'll try it." He was up and hurrying downstairs before he had finished talking. He flung open the front door, stepped through, and found himself staring at his companions, down the length of the second-floor lounge. "Well, there does seem to be some slight problem," he admitted blandly. "A mere technicality, though—we can always go out a window." He jerked aside the long drapes that covered the deep French windows set in one side wall of the lounge. He stopped suddenly.

"Hm-m-m," he said, "this is interesting—very."

"What is?" asked Bailey, joining him.

"This." The window stared directly into the dining room, instead of looking outdoors. Bailey stepped back to the corner where the lounge and the dining room joined the central room at ninety degrees.

"But that can't be," he protested, "that window is maybe fifteen, twenty feet from the dining room."

"Not in a tesseract," corrected Teal. "Watch." He opened the window and stepped through, talking back over his shoulder as he did so.

From the point of view of the Baileys he simply disappeared.

But not from his own viewpoint. It took him some seconds to catch his breath. Then he cautiously disentangled himself from the rosebush to which he had become almost irrevocably wedded, making a mental note the while never again to order landscaping which involved plants with thorns, and looked around him.

He was outside the house. The massive bulk of the ground-floor room thrust up beside him. Apparently he had fallen off the roof.

He dashed around the corner of the house, flung open the front door and hurried up the stairs. "Homer!" he called out, "Mrs. Bailey! I've found a way out!"

Bailey looked annoyed rather than pleased to see him. "What happened to you?"

"I fell out. I've been outside the house. You can do it just as easily —just step through those French windows. Mind the rosebush, though—we may have to build another stairway."

"How did you get back in?"

"Through the front door."

"Then we shall leave the same way. Come, my dear." Bailey set his hat firmly on his head and marched down the stairs, his wife on his arm.

Teal met them in the lounge. "I could have told you that wouldn't work," he announced. "Now here's what we have to do: As I see it, in a four-dimensional figure a three-dimensional man has two choices every time he crosses a line of juncture, like a wall or a threshold. Ordinarily he will make a ninety-degree turn through the fourth dimension, only he doesn't feel it with his three dimensions. Look." He stepped through the very window that he had fallen out of a moment before. Stepped through and arrived in the dining room, where he stood, still talking.

"I watched where I was going and arrived where I intended to." He stepped back into the lounge. "The time before I didn't watch and I moved on through normal space and fell out of the house. It must be a matter of subconscious orientation."

"I'd hate to depend on subconscious orientation when I step out for the morning paper."

"You won't have to; it'll become automatic. Now to get out of the house this time—Mrs. Bailey, if you will stand here with your back to the window, and jump backward, I'm pretty sure you will land in the garden."

Mrs. Bailey's face expressed her opinion of Teal and his ideas. "Homer Bailey," she said shrilly, "are you going to stand there and let him suggest such—"

"But Mrs. Bailey," Teal attempted to explain, "we can tie a rope on you and lower you down eas—"

"Forget it, Teal," Bailey cut him off brusquely. "We'll have to find a better way than that. Neither Mrs. Bailey nor I are fitted for jumping."

Teal was temporarily nonplused; there ensued a short silence. Bailey broke it with, "Did you hear that, Teal?"

"Hear what?"

"Someone talking off in the distance. D'you s'pose there could be someone else in the house, playing tricks on us, maybe?"

"Oh, not a chance. I've got the only key."

"But I'm sure of it," Mrs. Bailey confirmed. "I've heard them ever since we came in. Voices. Homer, I can't stand much more of this. Do something."

"Now, now, Mrs. Bailey," Teal soothed, "don't get upset. There can't be anyone else in the house, but I'll explore and make sure. Homer, you stay here with Mrs. Bailey and keep an eye on the rooms on this floor." He passed from the lounge into the ground-floor room and from there to the kitchen and on into the bedroom. This led him back to the lounge by a straight-line route, that is to say, by going straight ahead on the entire trip he returned to the place from which he started.

"Nobody around," he reported. "I opened all of the doors and windows as I went—all except this one." He stepped to the window opposite the one through which he had recently fallen and thrust back the drapes.

He saw a man with back toward him, four rooms away. Teal snatched open the French window and dived through it, shouting, "There he goes now! Stop, thief!"

The figure evidently heard him; it fled precipitately. Teal pursued, his gangling limbs stirred to unanimous activity, through drawing room, kitchen, dining room, lounge—room after room, yet in spite of Teal's best efforts he could not seem to cut down the four-room lead that the interloper had started with.

He saw the pursued jump awkwardly but actively over the low sill of a French window and in so doing knock off his hat. When he came up to the point where his quarry had lost his headgear, he stooped and picked it up, glad of an excuse to stop and catch his breath. He was back in the lounge.

"I guess he got away from me," he admitted. "Anyhow, here's his hat. Maybe we can identify him."

Bailey took the hat, looked at it, then snorted and slapped it on Teal's head. It fitted perfectly. Teal looked puzzled, took the hat off, and examined it. On the sweat band were the initials "Q.T." It was his own.

Slowly comprehension filtered through Teal's features. He went back to the French window and gazed down the series of rooms through which he had pursued the mysterious stranger. They saw him wave his arms semaphore fashion. "What are you doing?" asked Bailey.

"Come see." The two joined him and followed his stare with their own. Four rooms away they saw the backs of three figures, two male and one female. The taller, thinner of the men was waving his arms in a silly fashion.

Mrs. Bailey screamed and fainted again.

Some minutes later, when Mrs. Bailey had been resuscitated and somewhat composed, Bailey and Teal took stock. "Teal," said Bailey, "I won't waste any time blaming you; recriminations are useless and I'm sure you didn't plan for this to happen, but I suppose you realize we are in a pretty serious predicament. How are we going to get out of here? It looks now as if we would stay until we starve; every room leads into another room."

"Oh, it's not that bad. I got out once, you know."

"Yes, but you can't repeat it—you tried."

"Anyhow we haven't tried all the rooms. There's still the study."

"Oh, yes, the study. We went through there when we first came in, and didn't stop. Is it your idea that we might get out through its windows?"

"Don't get your hopes up. Mathematically, it ought to look into the four side rooms on this floor. Still we never opened the blinds; maybe we ought to look."

" 'Twon't do any harm anyhow. Dear, I think you had best just stay here and rest—"

"Be left alone in this horrible place? I should say not!" Mrs. Bailey was up off the couch where she had been recuperating even as she spoke.

They went upstairs. "This is the inside room, isn't it, Teal?" Bailey inquired as they passed through the master bedroom and climbed on up toward the study. "I mean it was the little cube in your diagram that was in the middle of the big cube, and completely surrounded."

"That's right," agreed Teal. "Well, let's have a look. I figure this window ought to give into the kitchen." He grasped the cords of Venetian blinds and pulled them.

It did not. Waves of vertigo shook them. Involuntarily they fell to the floor and grasped helplessly at the pattern on the rug to keep from falling. "Close it! Close it!" moaned Bailey.

Mastering in part a primitive atavistic fear, Teal worked his way back to the window and managed to release the screen. The window had looked *down* instead of *out*, down from a terrifying height.

Mrs. Bailey had fainted again.

Teal went back after more brandy while Bailey chafed her wrists. When she had recovered, Teal went cautiously to the window and raised the screen a crack. Bracing his knees, he studied the scene. He turned to Bailey. "Come look at this, Homer. See if you recognize it."

"You stay away from there, Homer Bailey!"

"Now, Matilda, I'll be careful." Bailey joined him and peered out.

"See up there? That's the Chrysler Building, sure as shooting. And there's the East River, and Brooklyn." They gazed straight down the sheer face of an enormously tall building. More than a thousand feet away a toy city, very much alive, was spread out before them. "As near as I can figure it out, we are looking down thc side of the Empire State Building from a point just above its tower."

"What is it? A mirage?"

"I don't think so—it's too perfect. I think space is folded over through the fourth dimension here and we are looking past the fold."

"You mean we aren't really seeing it?"

"No, we're seeing it all right. I don't know what would happen if we climbed out this window, but I for one don't want to try. But what a view! Oh, boy, what a view! Let's try the other windows."

They approached the next window more cautiously, and it was well that they did, for it was even more disconcerting, more reason-shaking, than the one looking down the gasping height of the skyscraper. It was a simple seascape, open ocean and blue sky—but the ocean was where the sky should have been, and contrariwise. This time they were somewhat braced for it, but they both felt seasickness about to overcome them at the sight of waves rolling overhead; they lowered the blind quickly without giving Mrs. Bailey a chance to be disturbed by it.

Teal looked at the third window. "Game to try it, Homer?"

"Hrrumph—well, we won't be satisfied if we don't. Take it easy." Teal lifted the blind a few inches. He saw nothing, and raised it a little more—still nothing. Slowly he raised it until the window was fully exposed. They gazed out at—nothing.

Nothing, nothing at all. What color is nothing? Don't be silly! What shape is it? Shape is an attribute of *something*. It had neither depth nor form. It had not even blackness. It was *nothing*.

Bailey chewed at his cigar. "Teal, what do you make of that?"

Teal's insouciance was shaken for the first time. "I don't know, Homer, I don't rightly know—but I think that window ought to be walled up." He stared at the lowered blind for a moment. "I think maybe we looked at a place where space *isn't*. We looked around a fourth-dimensional corner and there wasn't anything there." He rubbed his eyes. "I've got a headache."

They waited for a while before tackling the fourth window. Like an unopened letter, it might *not* contain bad news. The doubt left hope. Finally the suspense stretched too thin and Bailey pulled the cord himself, in the face of his wife's protests.

It was not so bad. A landscape stretched away from them, right side up, and on such a level that the study appeared to be a ground floor room. But it was distinctly unfriendly.

A hot, hot sun beat down from lemon-colored sky. The flat ground seemed burned a sterile, bleached brown and incapable of supporting life. Life there was, strange stunted trees that lifted knotted, twisted arms to the sky. Little clumps of spiky leaves grew on the outer extremities of these misshapen growths.

"Heavenly day," breathed Bailey, "where is that?"

Teal shook his head, his eyes troubled. "It beats me."

"It doesn't look like anything on Earth. It looks more like another planet—Mars, maybe."

"I wouldn't know. But, do you know, Homer, it might be worse than that, worse than another planet, I mean."

"Huh? What's that you say?"

"It might be clear out of our space entirely. I'm not sure that that is our sun at all. It seems too bright."

Mrs. Bailey had somewhat timidly joined them and now gazed out at the outré scene. "Homer," she said in a subdued voice, "those hideous trees—they frighten me."

He patted her hand.

Teal fumbled with the window catch.

"What are you doing?" Bailey demanded.

"I thought if I stuck my head out the window I might be able to look around and tell a bit more."

"Well—all right," Bailey grudged, "but be careful."

"I will." He opened the window a crack and sniffed. "The air is all right, at least." He threw it open wide.

His attention was diverted before he could carry out his plan. An uneasy tremor, like the first intimation of nausea, shivered the entire building for a long second, and was gone.

"Earthquake!" They all said it at once. Mrs. Bailey flung her arms around her husband's neck.

Teal gulped and recovered himself, saying, "It's all right, Mrs. Bailey. This house is perfectly safe. You know you can expect settling tremors after a shock like last night." He had just settled his features into an expression of reassurance when the second shock came. This one was no mild shimmy but the real sea-sick roll.

In every Californian, native born or grafted, there is a deep-rooted primitive reflex. An earthquake fills him with soul-shaking claustrophobia which impels him blindly to *get outdoors!* Model Boy Scouts will push aged grandmothers aside to obey it. It is a matter of record that Teal and Bailey landed on top of Mrs. Bailey. Therefore, she must have jumped through the window first. The order of precedence cannot be attributed to chivalry; it must be assumed that she was in readier position to spring.

They pulled themselves together, collected their wits a little, and rubbed sand from their eyes. Their first sensations were relief at feeling the solid sand of the desert land under them. Then Bailey noticed something that brought them to their feet and checked Mrs. Bailey from bursting into the speech that she had ready.

"Where's the house?"

It was gone. There was no sign of it at all. They stood in the center of flat desolation, the landscape they had seen from the window. But, aside from the tortured, twisted trees there was nothing to be seen but the yellow sky and the luminary overhead, whose furnacelike glare was already almost insufferable.

Bailey looked slowly around, then turned to the architect. "Well, Teal?" His voice was ominous.

Teal shrugged helplessly. "I wish I knew. I wish I could even be sure that we were on Earth."

"Well, we can't stand here. It's sure death if we do. Which direction?"

"Any, I guess. Let's keep a bearing on the sun."

They had trudged on for an undetermined distance when Mrs. Bailey demanded a rest. They stopped. Teal said in an aside to Bailey, "Any ideas?"

"No . . . no, none. Say, do you hear anything?"

Teal listened. "Maybe—unless it's my imagination."

"Sounds like an automobile. Say, it *is* an automobile!"

They came to the highway in less than another hundred yards. The automobile, when it arrived, proved to be an elderly, puffing light truck, driven by a rancher. He crunched to a stop at their hail. "We're stranded. Can you help us out?"

"Sure. Pile in."

"Where are you headed?"

"Los Angeles."

"Los Angeles? Say, where is this place?"

"Well, you're right in the middle of the Joshua-Tree National Forest."

The return was as dispiriting as the Retreat from Moscow. Mr. and Mrs. Bailey sat up in front with the driver while Teal bumped along in the body of the truck, and tried to protect his head from the sun. Bailey subsidized the friendly rancher to detour to the tesseract house, not because they wanted to see it again, but in order to pick up their car.

At last the rancher turned the corner that brought them back to where they had started. But the house was no longer there.

There was not even the ground floor room. It had vanished. The Baileys, interested in spite of themselves, poked around the foundations with Teal.

"Got any answers for this one, Teal?" asked Bailey.

"It must be that on that last shock it simply fell through into another section of space. I can see now that I should have anchored it at the foundations."

"That's not all you should have done."

"Well, I don't see that there is anything to get downhearted about. The house was insured, and we've learned an amazing lot. There are possibilities, man, possibilities! Why, right now I've got a great new revolutionary idea for a house—"

Teal ducked in time. He was always a man of action.

## RUSSELL MALONEY

# Inflexible Logic

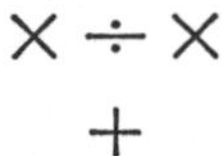

When the six chimpanzees came into his life, Mr. Bainbridge was thirty-eight years old. He was a bachelor and lived comfortably in a remote part of Connecticut, in a large old house with a carriage drive, a conservatory, a tennis court, and a well-selected library. His income was derived from impeccably situated real estate in New York City, and he spent it soberly, in a manner which could give offense to nobody. Once a year, late in April, his tennis court was resurfaced, and after that anybody in the neighborhood was welcome to use it; his monthly statement from Brentano's seldom ran below seventy-five dollars; every third year, in November, he turned in his old Cadillac coupé for a new one; he ordered his cigars, which were mild and rather moderately priced, in shipments of one thousand, from a tobacconist in Havana; because of the international situation he had canceled arrangements to travel abroad, and after due thought had decided to spend his traveling allowance on wines, which seemed likely to get scarcer and more expensive if the war lasted. On the whole, Mr. Bainbridge's life was deliberately, and not too unsuccessfully, modeled after that of an English country gentleman of the late eighteenth century, a gentleman interested in the arts and in the expansion of science, and so sure of himself that he didn't care if some people thought him eccentric.

Mr. Bainbridge had many friends in New York, and he spent several days of the month in the city, staying at his club and looking around. Sometimes he called up a girl and took her out to a theater and a night club. Sometimes he and a couple of classmates got a little tight and went to a prizefight. Mr. Bainbridge also looked in now and then at some of the conservative art galleries, and liked occasionally to go to a concert. And he liked cocktail parties, too, because of the fine footling conversation and the extraordinary number of pretty girls who had nothing else to do with the rest of their evening. It was at a New York cocktail party, however, that Mr. Bainbridge kept his preliminary appointment with doom. At one of the parties given by Hobie Packard, the stockbroker, he learned about the theory of the six chimpanzees.

It was almost six-forty. The people who had intended to have one drink and go had already gone, and the people who intended to stay were fortifying themselves with slightly dried canapés and talking animatedly. A group of stage and radio people had coagulated in one corner, near Packard's Capehart, and were wrangling about various methods of cheating the Collector of Internal Revenue. In another corner was a group of stockbrokers, talking about the greatest stockbroker of them all, Gauguin. Little Marcia Lupton was sitting with a young man, saying earnestly, "Do you really want to know what my greatest ambition is? I want to be myself," and Mr. Bainbridge smiled gently, thinking of the time Marcia had said that to him. Then he heard the voice of Bernard Weiss, the critic, saying, "Of course he wrote one good novel. It's not surprising. After all, we know that if six chimpanzees were set to work pounding six typewriters at random, they would, in a million years, write all the books in the British Museum."

Mr. Bainbridge drifted over to Weiss and was introduced to Weiss's companion, a Mr. Noble. "What's this about a million chimpanzees, Weiss?" he asked.

"Six chimpanzees," Mr. Weiss said. "It's an old cliché of the mathematicians. I thought everybody was told about it in school. Law of averages, you know, or maybe it's permutation and combination. The six chimps, just pounding away at the typewriter keys, would be bound to copy out all the books ever written by man. There are only so many possible combinations of letters and numerals, and they'd produce all of them—see? Of course they'd also turn out a mountain of gibberish, but they'd work the books in, too. All the books in the British Museum."

Mr. Bainbridge was delighted; this was the sort of talk he liked to hear when he came to New York. "Well, but look here," he said, just to keep up his part in the foolish conversation, "what if one of the chimpanzees finally did duplicate a book, right down to the last period, but left that off? Would that count?"

"I suppose not. Probably the chimpanzee would get around to doing the book again, and put the period in."

"What nonsense!" Mr. Noble cried.

"It may be nonsense, but Sir James Jeans believes it," Mr. Weiss said, huffily. "Jeans or Lancelot Hogben. I know I ran across it quite recently."

Mr. Bainbridge was impressed. He read quite a bit of popular science, and both Jeans and Hogben were in his library. "Is that so?" he murmured, no longer feeling frivolous. "Wonder if it has ever actually been tried? I mean, has anybody ever put six champanzees in a room with six typewriters and a lot of paper?"

Mr. Weiss glanced at Mr. Bainbridge's empty cocktail glass and said dryly, "Probably not."

Nine weeks later, on a winter evening, Mr. Bainbridge was sitting in his study with his friend James Mallard, an assistant professor of mathematics at New Haven. He was plainly nervous as he poured himself a drink and said, "Mallard, I've asked you to come here—brandy? cigar?—for a particular reason. You remember that I wrote you some time ago, asking your opinion of . . . of a certain mathematical hypothesis or supposition."

"Yes," Professor Mallard said, briskly. "I remember perfectly. About the six chimpanzees and the British Museum. And I told you it was a perfectly sound popularization of a principle known to every schoolboy who had studied the science of probabilities."

"Precisely," Mr. Bainbridge said. "Well, Mallard, I made up my mind. . . . It was not difficult for me, because I have, in spite of that fellow in the White House, been able to give something every year to the Museum of Natural History, and they were naturally glad to oblige me. . . . And after all, the only contribution a layman can make to the progress of science is to assist with the drudgery of experiment. . . . In short, I—"

"I suppose you're trying to tell me that you have procured six chimpanzees and set them to work at typewriters in order to see whether they will eventually write all the books in the British Museum. Is that it?"

"Yes, that's it," Mr. Bainbridge said. "What a mind you have, Mallard. Six fine young males, in perfect condition. I had a—I suppose you'd call it a dormitory—built out in back of the stable. The typewriters are in the conservatory. It's light and airy in there, and I moved most of the plants out. Mr. North, the man who owns the circus, very obligingly let me engage one of his best animal men. Really, it was no trouble at all."

Professor Mallard smiled indulgently. "After all, such a thing is not unheard of," he said. "I seem to remember that a man at some university put his graduate students to work flipping coins, to see if heads and tails came up an equal number of times. Of course they did."

Mr. Bainbridge looked at his friend very queerly. "Then you believe that any such principle of the science of probabilities will stand up under an actual test?"

"Certainly."

"You had better see for yourself." Mr. Bainbridge led Professor Mallard downstairs, along a corridor, through a disused music room, and into a large conservatory. The middle of the floor had been cleared of plants and was occupied by a row of six typewriter tables, each one supporting a hooded machine. At the left of each typewriter was a neat stack of yellow copy paper. Empty wastebaskets were under each table. The chairs were the unpadded, spring-backed kind favored by experienced stenographers. A large bunch of ripe bananas was hanging in one corner, and in another stood a Great Bear watercooler and a rack of Lily cups. Six piles of typescript, each about a foot high, were ranged along the wall on an improvised shelf. Mr. Bainbridge picked up one of the piles, which he could just conveniently lift, and set it on a table before Professor Mallard. "The output to date of Chimpanzee A, known as Bill," he said simply.

" '*Oliver Twist*, by Charles Dickens,' " Professor Mallard read out. He read the first and second pages of the manuscript, then feverishly leafed through to the end. "You mean to tell me," he said, "that this chimpanzee has written—"

"Word for word and comma for comma," said Mr. Bainbridge. "Young, my butler, and I took turns comparing it with the edition I own. Having finished *Oliver Twist*, Bill, is, as you see, starting the sociological works of Vilfredo Pareto, in Italian. At the rate he has been going, it should keep him busy for the rest of the month."

"And all the chimpanzees—" Professor Mallard was pale, and enunciated with difficulty—"they aren't all—"

"Oh, yes, all writing books which I have every reason to believe are in the British Museum. The prose of John Donne, some Anatole France, Conan Doyle, Galen, the collected plays of Somerset Maugham, Marcel Proust, the memoirs of the late Marie of Rumania, and a monograph by a Dr. Wiley on the marsh grasses of Maine and Massachusetts. I can sum it up for you, Mallard, by telling you that since I started this experiment, four weeks and some days ago, none of the chimpanzees has spoiled a single sheet of paper."

Professor Mallard straightened up, passed his handkerchief across his brow, and took a deep breath. "I apologize for my weakness," he said. "It was simply the sudden shock. No, looking at the thing scientifically—and I hope I am at least as capable of that as the next man —there is nothing marvelous about the situation. These chimpanzees, or a succession of similar teams of champanzees, would in a million years write all the books in the British Museum. I told you some time ago that I believed that statement. Why should my belief be altered by the fact that they produced some of the books at the very outset? After all, I should not be very much surprised if I tossed a coin a hundred times and it came up heads every time. I know that if I kcpt at it long cnough, thc ratio would reduce itself to an exact fifty pcr ccnt. Rcst assurcd, thcsc chimpanzees will begin to compose gibberish quite soon. It is bound to happen. Science tells us so. Meanwhile, I advise you to keep this experiment secret. Uninformed people might create a sensation if they knew."

"I will, indeed," Mr. Bainbridge said. "And I'm very grateful for your rational analysis. It reassures me. And now, before you go, you must hear the new Schnabel records that arrived today."

During the succeeding three months, Professor Mallard got into the habit of telephoning Mr. Bainbridge every Friday afternoon at five-thirty, immediately after leaving his seminar room. The Professor would say, "Well?" and Mr. Bainbridge would reply, "They're still at it, Mallard. Haven't spoiled a sheet of paper yet." If Mr. Bainbridge had to go out on Friday afternoon he would leave a written message with his butler, who would read it to Professor Mallard: "Mr. Bainbridge says we now have Trevelyan's *Life of Macaulay*, the Confessions of St. Augustine, *Vanity Fair*, part of Irving's *Life of George Washington*, the Book of the Dead, and some speeches delivered in Parliament in opposition to the Corn Laws, sir." Professor Mallard would reply, with a hint of a snarl in his voice, "Tell him to remember what I predicted," and hang up with a clash.

The eleventh Friday that Professor Mallard telephoned, Mr. Bainbridge said, "No change. I have had to store the bulk of the manuscript in the cellar. I would have burned it, except that it probably has some scientific value."

"How dare you talk of scientific value?" The voice from New Haven roared faintly in the receiver. "Scientific value! You—you—chimpanzee!" There were further inarticulate sputterings, and Mr. Bainbridge hung up with a disturbed expression. "I am afraid Mallard is overtaxing himself," he murmured.

Next day, however, he was pleasantly surprised. He was leafing through a manuscript that had been completed the previous day by Chimpanzee D, Corky. It was the complete diary of Samuel Pepys, and Mr. Bainbridge was chuckling over the naughty passages, which were omitted in his own edition, when Professor Mallard was shown into the room. "I have come to apologize for my outrageous conduct on the telephone yesterday," the Professor said.

"Please don't think of it any more. I know you have many things on your mind," Mr. Bainbridge said. "Would you like a drink?"

"A large whisky, straight, please," Professor Mallard said. "I got rather cold driving down. No change, I presume?"

"No, none. Chimpanzee F, Dinty, is just finishing John Florio's translation of Montaigne's essays, but there is no other news of interest."

Professor Mallard squared his shoulders and tossed off his drink in one astonishing gulp. "I should like to see them at work," he said. "Would I disturb them, do you think?"

"Not at all. As a matter of fact, I usually look in on them around this time of day. Dinty may have finished his Montaigne by now, and it is always interesting to see them start a new work. I would have thought that they would continue on the same sheet of paper, but they don't, you know. Always a fresh sheet, and the title in capitals."

Professor Mallard, without apology, poured another drink and slugged it down. "Lead on," he said.

It was dusk in the conservatory, and the chimpanzees were typing by the light of student lamps clamped to their desks. The keeper lounged in a corner, eating a banana and reading *Billboard*. "You might as well take an hour or so off," Mr. Bainbridge said. The man left.

Professor Mallard, who had not taken off his overcoat, stood with his hands in his pockets, looking at the busy chimpanzees, "I wonder if you know, Bainbridge, that the science of probabilities takes everything into account," he said, in a queer, tight voice. "It is certainly almost beyond the bounds of credibility that these chimpanzees should write books without a single error, but that abnormality may be corrected by—*these!*" He took his hands from his pockets, and each one held a .38 revolver. "Stand back out of harm's way!" he shouted.

"Mallard! Stop it!" The revolvers barked, first the right hand, then the left, then the right. Two chimpanzees fell, and a third reeled into a corner. Mr. Bainbridge seized his friend's arm and wrested one of the weapons from him.

"Now I am armed, too, Mallard, and I advise you to stop!" he cried. Professor Mallard's answer was to draw a bead on Chimpanzee E and shoot him dead. Mr. Bainbridge made a rush, and Professor Mallard fired at him. Mr. Bainbridge, in his quick death agony, tightened his finger on the trigger of his revolver. It went off, and Professor Mallard went down. On his hands and knees he fired at the two chimpanzees which were still unhurt, and then collapsed.

There was nobody to hear his last words. "The human equation . . . always the enemy of science . . ." he panted. "This time . . . vice versa . . . I, a mere mortal, . . . savior of science . . . deserve a Nobel . . ."

When the old butler came running into the conservatory to investigate the noises, his eyes were met by a truly appalling sight. The student lamps were shattered, but a newly risen moon shone in through the conservatory windows on the corpses of the two gentlemen, each clutching a smoking revolver. Five of the chimpanzees were dead. The sixth was Chimpanzee F. His right arm disabled, obviously bleeding to death, he was slumped before his typewriter. Painfully, with his left hand, he took from the machine the completed last page of Florio's Montaigne. Groping for a fresh sheet, he inserted it, and typed with one finger, "UNCLE TOM'S CABIN, by Harriet Beecher Stowe. Chapte . . ." Then he, too, was dead.

MARTIN GARDNER

# No-Sided Professor

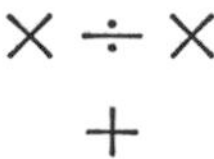

DOLORES—a tall, black-haired striptease at Chicago's Purple Hat Club—stood in the center of the dance floor and began the slow gyrations of her Cleopatra number, accompanied by soft Egyptian music from the Purple Hatters. The room was dark except for a shaft of emerald light that played over her filmy Egyptian costume and smooth, voluptuous limbs.

A veil draped about her head and shoulders was the first to be removed. Dolores was in the act of letting it drift gracefully to the floor when suddenly a sound like the firing of a shotgun came from somewhere above and the nude body of a large man dropped head first from the ceiling. He caught the veil in mid-air with his chin and pinned it to the floor with a dull thump.

Pandemonium reigned.

Jake Bowers, the master of ceremonies, yelled for lights and tried to keep back the crowd. The club's manager, who had been standing by the orchestra watching the floor show, threw a tablecloth over the crumpled figure and rolled it over on its back.

The man was breathing heavily, apparently knocked unconscious by the blow on his chin, but otherwise unharmed. He was well over fifty, with a short, neatly trimmed red beard and mustache, and a completely bald head. He was built like a professional wrestler.

With considerable difficulty three waiters succeeded in transporting him to the manager's private office in the back, leaving a roomful of bewildered, near-hysterical men and women gaping at the ceiling and each other, and arguing heatedly about the angle and manner of the man's fall. The only hypothesis with even a slight suggestion of sanity was that he had been tossed high into the air from somewhere on the side of the dance floor. But no one saw the tossing. The police were called.

Meanwhile, in the back office the bearded man recovered consciousness. He insisted that he was Dr. Stanislaw Slapenarski, professor of mathematics at the University of Warsaw, and at present a visiting lecturer at the University of Chicago.

Before continuing this curious narrative, I must pause to confess that I was not an eyewitness of the episode just described, having based my account on interviews with the master of ceremonies and several waiters. However, I did participate in a chain of remarkable events which culminated in the professor's unprecedented appearance.

These events began several hours earlier when members of the Moebius Society gathered for their annual banquet in one of the private dining rooms on the second floor of the Purple Hat Club. The Moebius Society is a small, obscure Chicago organization of mathematicians working in the field of topology, one of the youngest and most mysterious of the newer branches of transformation mathematics. To make clear what happened during the evening, it will be necessary at this point to give a brief description of the subject matter of topology.

Topology is difficult to define in nontechnical terms. One way to put it is to say that topology studies the mathematical properties of an object which remain constant regardless of how the object is distorted.

Picture in your mind a doughnut made of soft pliable rubber that can be twisted and stretched as far as you like in any direction. No matter how much this rubber doughnut is distorted (or "transformed" as mathematicians prefer to say), certain properties of the doughnut will remain unchanged. For example, it will always retain a hole. In topology the doughnut shape is called a "torus." A soda straw is merely an elongated torus, so—from a topological point of view—a doughnut and a soda straw are identical figures.

Topology is completely disinterested in quantitative measurements.

It is concerned only with basic properties of shape which are unchanged throughout the most radical distortions possible without breaking off pieces of the object and sticking them on again at other spots. If this breaking off were permitted, an object of a given structure could be transformed into an object of any other type of structure, and all original properties would be lost. If the reader will reflect a moment he will soon realize that topology studies the most primitive and fundamental mathematical properties that an object can possess.[1]

A sample problem in topology may be helpful. Imagine a torus (doughnut) surface made of thin rubber like an inner tube. Now imagine a small hole in the side of this torus. Is it possible to turn the torus inside out through this hole, as you might turn a balloon inside out? This is not an easy problem to solve in the imagination.

Although many mathematicians of the eighteenth century wrestled with isolated topological problems, one of the first systematic works in the field was done by August Ferdinand Moebius, a German astronomer who taught at the University of Leipzig during the first half of the last century. Until the time of Moebius it was believed that any surface, such as a piece of paper, had two sides. It was the German astronomer who made the disconcerting discovery that if you take a strip of paper, give it a single half-twist, then paste the ends together, the result is a "unilateral" surface—a surface with only *one* side!

If you will trouble to make such a strip (known to topologists as the "Moebius surface") and examine it carefully, you will soon discover that the strip actually does consist of only one continuous side and of one continuous edge.

It is hard to believe at first that such a strip can exist, but there it is—a visible, tangible thing that can be constructed in a moment. And it has the indisputable property of one-sidedness, a property it

[1] The reader who is interested in obtaining a clearer picture of this new mathematics will find excellent articles on topology in the *Encyclopaedia Britannica* (Fourteenth Edition) under *Analysis Situs;* and under *Analysis Situs* in the *Encyclopedia Americana.* There also are readable chapters on elementary topology in two recent books—*Mathematics and the Imagination* by Kasner and Newman, and *What Is Mathematics?* by Courant and Robbins. Slapenarski's published work has not yet been translated from the Polish.

cannot lose no matter how much it is stretched or how it is distorted.[2]

But back to the story. As an instructor in mathematics at the University of Chicago with a doctor's thesis in topology to my credit, I had little difficulty in securing admittance into the Moebius Society. Our membership was small—only twenty-six men, most of them Chicago topologists but a few from universities in neighboring towns.

We held regular monthly meetings, rather academic in character, and once a year on November 17 (the anniversary of Moebius' birth) we arranged a banquet at which an outstanding topologist was brought to the city to act as a guest speaker.

The banquet always had its less serious aspects, usually in the form of special entertainment. But this year our funds were low and we decided to hold the celebration at the Purple Hat where the cost of the dinner would not be too great and where we could enjoy the floor show after the lecture. We were fortunate in having been able to obtain as our guest the distinguished Professor Slapenarski, universally acknowledged as the world's leading topologist and one of the greatest mathematical minds of the century.

Dr. Slapenarski had been in the city several weeks giving a series of lectures at the University of Chicago on the topological aspects of Einstein's theory of space. As a result of my contacts with him at the university, we became good friends and I had been asked to introduce him at the dinner.

We rode to the Purple Hat together in a taxi, and on the way I begged him to give me some inkling of the content of his address. But he only smiled inscrutably and told me, in his thick Polish accent, to wait and see. He had announced his topic as "The No-Sided Surface"—a topic which had aroused such speculation among our members that Dr. Robert Simpson of the University of Wisconsin

[2] The Moebius strip has many terrifying properties. For example, if you cut the strip in half lengthwise, cutting down the center all the way around, the result is not two strips, as might be expected, but one single large strip. But if you begin cutting a third of the way from the side, cutting *twice* around the strip, the result is one large and one small strip, interlocked. The smaller strip can then be cut in half to yield a single large strip, still interlocked with the other large strip. These weird properties are the basis of an old magic trick with cloth, known to the conjuring profession as the "Afghan bands."

wrote he was coming to the dinner, the first meeting that he had attended in over a year.[3]

Dr. Simpson is the outstanding authority on topology in the Middle West and the author of several important papers on topology and nuclear physics in which he vigorously attacks several of Slapenarski's major axioms.

The Polish professor and I arrived a little late. After introducing him to Simpson, then to our other members, we took our seats at the table and I called Slapenarski's attention to our tradition of brightening the banquet with little topological touches. For instance, our napkin rings were silver-plated Moebius strips. Doughnuts were provided with the coffee, and the coffee itself was contained in specially designed cups made in the shape of "Klein's bottle."[4]

After the meal we were served Ballantine's ale, because of the curious trade-mark,[5] and pretzels in the shapes of the two basic "trefoil" knots.[6] Slapenarski was much amused by these details and even made several suggestions for additional topological curiosities, but the suggestions are too complex to explain here.

After my brief introduction, the Polish doctor stood up, acknowledged the applause with a smile, and cleared his throat. The room instantly became silent. The reader is already familiar with the professor's appearance—his portly frame, reddish beard, and polished

[3] Dr. Simpson later confided to me that he had attended the dinner not to hear Slapenarski but to see Dolores.

[4] Named after Felix Klein, a brilliant German mathematician, Klein's bottle is a completely closed surface, like the surface of a globe, but without inside or outside. It is unilateral like a Moebius strip, but unlike the strip it has no edges. It can be bisected in such a way that each half becomes a Moebius surface. It will hold a liquid. Nothing frightful happens to the liquid.

[5] This trade-mark is a topological manifold of great interest. Although the three rings are interlocked, no *two* rings are interlocked. In other words, if any one of the rings is removed, the other two rings are completely free of each other. Yet the three together cannot be separated.

[6] The trefoil knot is the simplest form of knot that can be tied in a closed curve. It exists in two forms, one a mirror image of the other. Although the two forms are topologically identical, it is impossible to transform one into the other by distortion, an upsetting fact that has caused topologists considerable embarrassment. The study of the properties of knots forms an important branch of topology, though very little is understood as yet about even the simplest knots.

pate—but it should be added that there was something in the expression of his face that suggested that he had matters of considerable import to disclose to us.

It would be impossible to give with any fullness the substance of Slapenarski's brilliant, highly technical address. But the gist of it was this. Ten years ago, he said, he had been impressed by a statement of Moebius, in one of his lesser known treatises, that there was no theoretical reason why a surface could not lose *both* its sides—to become in other words, a "nonlateral" surface.

Of course, the professor explained, such a surface was impossible to imagine, but so is the square root of minus one or the hypercube of fourth-dimensional geometry. That a concept is inconceivable has long ago been recognized as no basis for denying either its validity or usefulness in mathematics and modern physics.

We must remember, he added, that even the one-sided surface is inconceivable to anyone who has not seen and handled a Moebius strip. And many persons, with well-developed mathematical imaginations, are unable to understand how such a strip can exist even when they have one in hand.

I glanced at Dr. Simpson and thought I detected a skeptical smile curving the corners of his mouth.

Slapenarski continued. For many years, he said, he had been engaged in a tireless quest for a no-sided surface. On the basis of analogy with known types of surfaces he had been able to analyze many of the properties of the no-sided surface. Finally one day—and he paused here for dramatic emphasis, sweeping his bright little eyes across the motionless faces of his listeners—he had actually succeeded in constructing a no-sided surface.

His words were like an electric impulse that transmitted itself around the table. Everyone gave a sudden start and shifted his position and looked at his neighbor with raised eyebrows. I noticed that Simpson was shaking his head vigorously. When the speaker walked to the end of the room where a blackboard had been placed, Simpson bent his head and whispered to the man on his left, "It's sheer nonsense. Either Slappy has gone completely mad or he's playing a deliberate prank on all of us."

I think it had occurred to the others also that the lecture was a hoax because I noticed several were smiling to themselves while the professor chalked some elaborate diagrams on the blackboard.

After a somewhat involved discussion of the diagrams (which I was wholly unable to follow) the professor announced that he would conclude his lecture by constructing one of the simpler forms of the no-sided surface. By now we were all grinning at each other. Dr. Simpson's face had more of a smirk than a grin.

Slapenarski produced from his coat pocket a sheet of pale blue paper, a small pair of scissors, and a tube of paste. He cut the paper into a figure that had a striking resemblance, I thought, to a paper doll. There were five projecting strips or appendages that resembled a head and four limbs. Then he folded and pasted the sheet carefully. It was an intricate procedure. Strips went over and under each other in an odd fashion until finally only two ends projected. Dr. Slapenarski then applied a dab of paste to one of these ends.

"Gentlemen," he said, holding up the twisted blue construction and turning it about for all to see, "you are about to witness the first public demonstration of the Slapenarski surface."

So saying, he pressed one of the projecting ends against the other.

There was a loud pop, like the bursting of a light bulb, and the paper figure vanished in his hands!

For a moment we were too stunned to move, then with one accord we broke into laughter and applause.

We were convinced, of course, that we were the victims of an elaborate joke. But it had been beautifully executed. I assumed, as did the others, that we had witnessed an ingenious chemical trick with paper—paper treated so it could be ignited by friction or some similar method and caused to explode without leaving an ash.

But I noticed that the professor seemed disconcerted by the laughter, and his face was beginning to turn the color of his beard. He smiled in an embarrassed way and sat down. The applause subsided slowly.

Falling in with the preposterous mood of the evening we all clustered around him and congratulated him warmly on his remarkable discovery. Then the man in charge of arrangements reminded us that a table had been reserved below so those interested in remaining could enjoy some drinks and see the floor show.

The room gradually cleared of everyone except Slapenarski, Simpson, and myself. The two famous topologists were standing in front of the blackboard. Simpson was smiling broadly and gesturing toward one of the diagrams.

"The fallacy in your proof was beautifully concealed, Doctor," he said. "I wonder if any of the others caught it."

The Polish mathematician was not amused.

"There is no fallacy in my proof," he said impatiently.

"Oh, come now, Doctor," Simpson said. "Of course there's a fallacy." Still smiling, he touched a corner of the diagram with his thumb. "These lines can't possibly intersect within the manifold. The intersection is somewhere out here." He waved his hand off to the right.

Slapenarski's face was growing red again.

"I tell you there is no fallacy," he repeated, his voice rising. Then slowly, speaking his words carefully and explosively, he went over the proof once more, rapping the blackboard at intervals with his knuckles.

Simpson listened gravely, and at one point interrupted with an objection. The objection was answered. A moment later he raised a second objection. The second objection was answered. I stood aside without saying anything. The discussion was too far above my head.

Then they began to raise their voices. I have already spoken of Simpson's long-standing controversy with Slapenarski over several basic topological axioms. Some of these axioms were now being brought into the argument.

"But I tell you the transformation is *not* bicontinuous and therefore the two sets cannot be homeomorphic," Simpson shouted.

The veins on the Polish mathematician's temples were standing out in sharp relief. "Then suppose you explain to me why my manifold vanished," he yelled back.

"It was nothing but a cheap conjuring trick," snorted Simpson. "I don't know how it worked and I don't care, but it certainly wasn't because the manifold became nonlateral."

"Oh it wasn't, wasn't it?" Slapenarski said between his teeth. Before I had a chance to intervene he had sent his huge fist crashing into the jaw of Dr. Simpson. The Wisconsin professor groaned and dropped to the floor. Slapenarski turned and glared at me wildly.

"Get back, young man," he said. As he outweighed me by at least one hundred pounds, I got back.

Then I watched in horror what was taking place. With insane fury still flaming on his face, Slapenarski had knelt beside the limp body and was twisting the arms and legs into fantastic knots. He was, in fact, folding the Wisconsin topologist as he had folded his piece of paper! Suddenly there was a small explosion, like the backfire of a car, and under the Polish mathematician's hands lay the collapsed clothing of Dr. Simpson.

Simpson had become a nonlateral surface.

Slapenarski stood up, breathing with difficulty and holding in his hands a tweed coat with vest, shirt, and underwear top inside. He opened his hands and let the garments fall on top of the clothing on the floor. Great drops of perspiration rolled down his face. He muttered in Polish, then beat his fists against his forehead.

I recovered enough presence of mind to move to the entrance of the room, and lock the door. When I spoke my voice sounded weak. "Can he . . . be brought back?"

"I do not know, I do not know," Slapenarski wailed. "I have only begun the study of these surfaces—only just begun. I have no way of knowing where he is. Undoubtedly it is one of the higher dimensions, probably one of the odd-numbered ones. God knows which one."

Then he grabbed me suddenly by my coat lapels and shook me so violently that a bridge on my upper teeth came loose. "I must go to him," he said. "It is the least I can do—the very least."

He sat down on the floor and began interweaving arms and legs. "Do not stand there like an idiot!" he yelled. "Here—some assistance."

I adjusted my bridge, then helped him twist his right arm under his left leg and back around his head until he was able to grip his right ear. Then his left arm had to be twisted in a somewhat similar fashion. "Over, not under," he shouted. It was with difficulty that I was able to force his left hand close enough to his face so he could grasp his nose.

There was another explosive noise, much louder than the sound made by Simpson, and a sudden blast of cold wind across my face. When I opened my eyes I saw the second heap of crumpled clothing on the floor.

While I was staring stupidly at the two piles of clothing there was a muffled sort of "pfft" sound behind me. I turned and saw Simpson standing near the wall, naked and shivering. His face was white. Then his knees buckled and he sank to the floor. There were vivid red marks at various places where his limbs had been pressed tightly against each other.

I stumbled to the door, unlocked it, and started down the stairway after a strong drink—for myself. I became conscious of a violent hubbub on the dance floor. Slapenarski had, a few moments earlier, completed his sensational dive.

In a back room below I found the other members of the Moebius Society and various officials of the Purple Hat Club in noisy, incoherent debate. Slapenarski was sitting in a chair with a tablecloth wrapped around him and holding a handkerchief filled with ice cubes against the side of his jaw.

"Simpson is back," I said. "He fainted but I think he's okay."

"Thank heavens," Slapenarski mumbled.

The officials and patrons of the Purple Hat never understood, of course, what happened that wild night, and our attempts to explain made matters worse. The police arrived, adding to the confusion.

We finally got the two professors dressed and on their feet, and made an escape by promising to return the following day with our lawyers. The manager seemed to think the club had been the victim of an outlandish plot, and threatened to sue for damages against what he called the club's "refined reputation." As it turned out, the incident proved to be magnificent word-of-mouth advertising and eventually the club dropped the case. The papers heard the story, of course, but promptly dismissed it as an uncouth publicity stunt cooked up by Phanstiehl, the Purple Hat's press agent.

Simpson was unhurt, but Slapenarski's jaw had been broken. I took him to Billings Hospital, near the university, and in his hospital room late that night he told me what he thought had happened. Apparently Simpson had entered a higher dimension (very likely the fifth) on level ground.

When he recovered consciousness he unhooked himself and immediately reappeared as a normal three-dimensional torus with outside and inside surfaces. But Slapenarski had worse luck. He had landed on some sort of slope. There was nothing to see—only a gray, undifferentiated fog on all sides—but he had the distinct sensation of rolling down a hill.

He tried to keep a grip on his nose but was unable to maintain it. His right hand slipped free before he reached the bottom of the incline. As a result, he unfolded himself and tumbled back into three-dimensional space and the middle of Dolores' Egyptian routine.

At any rate that was the way Slapenarski had it figured out.

He was several weeks in the hospital, refusing to see anyone until the day of his release, when I accompanied him to the Union Station. He caught a train to New York and I never saw him again. He died a few months later of a heart attack in Warsaw. At present Dr. Simpson is in correspondence with his widow in an attempt to obtain his notes on nonlateral surfaces.

Whether these notes will or will not be intelligible to American topologists (assuming we can obtain them) remains to be seen. We have made numerous experiments with folded paper, but so far have produced only commonplace bilateral and unilateral surfaces. Although it was I who helped Slapenarski fold himself, the excitement of the moment apparently erased the details from my mind.

But I shall never forget one remark the great topologist made to me the night of his accident, just before I left him at the hospital.

"It was fortunate," he said, "that both Simpson and I released our right hand before the left."

"Why?" I asked.

Slapenarski shuddered.

"We would have been inside out," he said.

ARTHUR C. CLARKE

# Superiority

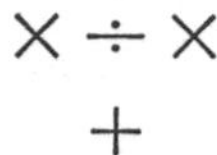

IN MAKING THIS STATEMENT—which I do of my own free will—I wish first to make it perfectly clear that I am not in any way trying to gain sympathy, nor do I expect any mitigation of whatever sentence the Court may pronounce. I am writing this in an attempt to refute some of the lying reports published in the papers I have been allowed to see, and broadcast over the prison radio. These have given an entirely false picture of the true cause of our defeat, and as the leader of my race's armed forces at the cessation of hostilities I feel it my duty to protest against such libels upon those who served under me.

I also hope that this statement may explain the reasons for the application I have twice made to the Court, and will now induce it to grant a favor for which I can see no possible grounds of refusal.

The ultimate cause of our failure was a simple one: despite all statements to the contrary, it was not due to lack of bravery on the part of our men, or to any fault of the Fleet's. We were defeated by one thing only—by the inferior science of our enemies. I repeat—by the *inferior* science of our enemies.

When the war opened we had no doubts of our ultimate victory. The combined fleets of our allies greatly exceeded in number and armament those which the enemy could muster against us, and in almost all branches of military science we were their superiors. We were sure that we could maintain this superiority. Our belief proved, alas, to be only too well-founded.

At the opening of the war our main weapons were the long-range homing torpedo, dirigible ball-lightning and the various modifications of the Klydon beam. Every unit of the Fleet was equipped with these, and though the enemy possessed similar weapons their installations were generally of lesser power. Moreover, we had behind us a far greater military Research Organization, and with this initial advantage we could not possibly lose.

The campaign proceeded according to plan until the Battle of the Five Suns. We won this, of course, but the opposition proved stronger than we had expected. It was realized that victory might be more difficult, and more delayed, than had first been imagined. A conference of supreme commanders was therefore called to discuss our future strategy.

Present for the first time at one of our war conferences was Professor General Norden, the new Chief of the Research Staff, who had just been appointed to fill the gap left by the death of Malvar, our greatest scientist. Malvar's leadership had been responsible, more than any other single factor, for the efficiency and power of our weapons. His loss was a very serious blow, but no one doubted the brilliance of his successor—though many of us disputed the wisdom of appointing a theoretical scientist to fill a post of such vital importance. But we had been overruled.

I can well remember the impression Norden made at that conference. The military advisers were worried, and as usual turned to the scientists to help. Would it be possible to improve our existing weapons, they asked, so that our present advantage could be increased still further?

Norden's reply was quite unexpected. Malvar had often been asked such a question—and he had always done what we requested.

"Frankly, gentlemen," said Norden, "I doubt it. Our existing weapons have practically reached finality. I don't wish to criticize my predecessor, or the excellent work done by the Research Staff in the last few generations, but do you realize that there has been no basic change in armaments for over a century? It is, I am afraid, the result of a tradition that has become conservative. For too long, the Research Staff has devoted itself to perfecting old weapons instead of developing new ones. It is fortunate for us that our opponents have been no wiser; we cannot assume that this will always be so."

Norden's words left an uncomfortable impression, as he had no doubt intended. He quickly pressed home the attack.

"What we want are *new* weapons—weapons totally different from any that have been employed before. Such weapons can be made; it will take time, of course, but since assuming charge I have replaced some of the older scientists by young men and have directed research into several unexplored fields which show great promise. I believe, in fact, that a revolution in warfare may soon be upon us."

We were skeptical. There was a bombastic tone in Norden's voice that made us suspicious of his claims. We did not know, then, that he never promised anything that he had not already almost perfected in the laboratory. *In the laboratory*—that was the operative phrase.

Norden proved his case less than a month later, when he demonstrated the Sphere of Annihilation, which produced complete disintegration of matter over a radius of several hundred meters. We were intoxicated by the power of the new weapon, and were quite prepared to overlook one fundamental defect—the fact that it *was* a sphere and hence destroyed its rather complicated generating equipment at the instant of formation. This meant, of course, that it could not be used on warships but only on guided missiles, and a great program was started to convert all homing torpedoes to carry the new weapon. For the time being all further offensives were suspended.

We realize now that this was our first mistake. I still think that it was a natural one, for it seemed to us then that all our existing weapons had become obsolete overnight, and we already regarded them almost as primitive survivals. What we did not appreciate was the magnitude of the task we were attempting, and the length of time it would take to get the revolutionary super-weapon into battle. Nothing like this had happened for a hundred years and we had no previous experience to guide us.

The conversion problem proved far more difficult than anticipated. A new class of torpedo had to be designed, as the standard mark was too small. This meant in turn that only the larger ships could launch the weapon, but we were prepared to accept this penalty. After six months, the heavy units of the Fleet were being equipped with the Sphere. Training maneuvers and tests had shown that it was operating satisfactorily and we were ready to take it into action. Norden was already being hailed as the architect of victory, and had half promised even more spectacular weapons.

Then two things happened. One of our battleships disappeared completely on a training flight, and an investigation showed that under certain conditions the ship's long-range radar could trigger the Sphere immediately it had been launched. The modification needed to overcome this defect was trivial, but it caused a delay of another month and was the source of much bad feeling between the naval staff and the scientists. We were ready for action again—when Norden announced that the radius of effectiveness of the Sphere had now been increased by ten, thus multiplying by a thousand the chances of destroying an enemy ship.

So the modifications started all over again, but everyone agreed that the delay would be worth it. Meanwhile, however, the enemy had been emboldened by the absence of further attacks and had made an unexpected onslaught. Our ships were short of torpedoes, since none had been coming from the factories, and were forced to retire. So we lost the systems of Kyrane and Floranus, and the planetary fortress of Rhamsandron.

It was an annoying but not a serious blow, for the recaptured systems had been unfriendly and difficult to administer. We had no doubt that we could restore the position in the near future, as soon as the new weapon became operational.

These hopes were only partially fulfilled. When we renewed our offensive, we had to do so with fewer of the Spheres of Annihilation than had been planned, and this was one reason for our limited success. The other reason was more serious.

While we had been equipping as many of our ships as we could with the irresistible weapon, the enemy had been building feverishly. His ships were of the old pattern, with the old weapons—but they now outnumbered ours. When we went into action, we found that the numbers ranged against us were often 100 per cent greater than expected, causing target confusion among the automatic weapons and resulting in higher losses than anticipated. The enemy losses were higher still, for once a Sphere had reached its objective, destruction was certain, but the balance had not swung as far in our favor as we had hoped.

Moreover, while the main fleets had been engaged, the enemy had launched a daring attack on the lightly held systems of Eriston, Duranus, Carmanidor and Pharanidon—recapturing them all. We were thus faced with a threat only fifty light-years from our home planets.

There was much recrimination at the next meeting of the supreme commanders. Most of the complaints were addressed to Norden—Grand Admiral Taxaris in particular maintaining that thanks to our admittedly irresistible weapon we were now considerably worse off than before. We should, he claimed, have continued to build conventional ships, thus preventing the loss of our numerical superiority.

Norden was equally angry and called the naval staff ungrateful bunglers. But I could tell that he was worried—as indeed we all were—by the unexpected turn of events. He hinted that there might be a speedy way of remedying the situation.

We now know that Research had been working on the Battle Analyzer for many years, but at the time it came as a revelation to us and perhaps we were too easily swept off our feet. Norden's argument, also, was seductively convincing. What did it matter, he said, if the enemy had twice as many ships as we—if the efficiency of ours could be doubled or even trebled? For decades the limiting factor in warfare had been not mechanical but biological—it had become more and more difficult for any single mind, or group of minds, to cope with the rapidly changing complexities of battle in three-dimensional space. Norden's mathematicians had analyzed some of the classic engagements of the past, and had shown that even when we had been victorious we had often operated our units at much less than half of their theoretical efficiency.

The Battle Analyzer would change all this by replacing operations staff with electronic calculators. The idea was not new, in theory, but until now it had been no more than a utopian dream. Many of us found it difficult to believe that it was still anything but a dream; after we had run through several very complex dummy battles, however, we were convinced.

It was decided to install the Analyzer in four of our heaviest ships, so that each of the main fleets could be equipped with one. At this stage, the trouble began—though we did not know it until later.

The Analyzer contained just short of a million vacuum tubes and needed a team of five hundred technicians to maintain and operate it. It was quite impossible to accommodate the extra staff aboard a battleship, so each of the four units had to be accompanied by a converted liner to carry the technicians not on duty. Installation was also a very slow and tedious business, but by gigantic efforts it was completed in six months.

Then, to our dismay, we were confronted by another crisis. Nearly five thousand highly skilled men had been selected to service the Analyzers and had been given an intensive course at the Technical Training Schools. At the end of seven months, 10 per cent of them had had nervous breakdowns and only 40 per cent had qualified.

Once again, everyone started to blame everyone else. Norden, of course, said that the Research Staff could not be held responsible, and so incurred the enmity of the Personnel and Training Commands. It was finally decided that the only thing to do was to use two instead of four Analyzers and to bring the others into action as soon as men could be trained. There was little time to lose, for the enemy was still on the offensive and his morale was rising.

The first Analyzer fleet was ordered to recapture the system of Eriston. On the way, by one of the hazards of war, the liner carrying the technicians was struck by a roving mine. A warship would have survived, but the liner with its irreplaceable cargo was totally destroyed. So the operation had to be abandoned.

The other expedition was, at first, more successful. There was no doubt at all that the Analyzer fulfilled its designers' claims, and the enemy was heavily defeated in the first engagements. He withdrew, leaving us in possession of Saphran, Leucon and Hexanerax. But his Intelligence staff must have noted the change in our tactics and the inexplicable presence of a liner in the heart of our battlefleet. It must have noted, also, that our first fleet had been accompanied by a similar ship—and had withdrawn when it had been destroyed.

In the next engagement, the enemy used his superior numbers to launch an overwhelming attack on the Analyzer ship and its unarmed consort. The attack was made without regard to losses—both ships were, of course, very heavily protected—and it succeeded. The result was the virtual decapitation of the fleet, since an effectual transfer to the old operational methods proved impossible. We disengaged under heavy fire, and so lost all our gains and also the systems of Lorymia, Ismarnus, Beronis, Alphanidon and Sideneus.

At this stage, Grand Admiral Taxaris expressed his disapproval of Norden by committing suicide, and I assumed supreme command.

The situation was now both serious and infuriating. With stubborn conservatism and complete lack of imagination, the enemy continued to advance with his old-fashioned and inefficient but now vastly more numerous ships. It was galling to realize that if we had only continued building, without seeking new weapons, we would have been in a far more advantageous position. There were many acrimonious conferences at which Norden defended the scientists while everyone else blamed them for all that had happened. The difficulty was that Norden had proved every one of his claims: he had a perfect excuse for all the disasters that had occurred. And we could not now turn back—the search for an irresistible weapon must go on. At first it had been a luxury that would shorten the war. Now it was a necessity if we were to end it victoriously.

We were on the defensive, and so was Norden. He was more than ever determined to re-establish his prestige and that of the Research Staff. But we had been twice disappointed, and would not make the same mistake again. No doubt Norden's twenty thousand scientists would produce many further weapons; we would remain unimpressed.

We were wrong. The final weapon was something so fantastic that even now it seems difficult to believe that it ever existed. Its innocent, noncommittal name—the Exponential Field—gave no hint of its real potentialities. Some of Norden's mathematicians had discovered it during a piece of entirely theoretical research into the properties of space, and to everyone's great surprise their results were found to be physically realizable.

It seems very difficult to explain the operation of the Field to the layman. According to the technical description, it "produces an exponential condition of space, so that a finite distance in normal, linear space may become infinite in pseudo-space." Norden gave an analogy which some of us found useful. It was as if one took a flat disc of rubber—representing a region of normal space—and then pulled its center out to infinity. The circumference of the disc would be unaltered—but its "diameter" would be infinite. That was the sort of thing the generator of the field did to the space around it.

As an example, suppose that a ship carrying the generator was surrounded by a ring of hostile machines. If it switched on the Field, *each* of the enemy ships would think that it—and the ships on the far side of the circle—had suddenly receded into nothingness. Yet the circumference of the circle would be the same as before; only the journey to the center would be of infinite duration, for as one proceeded, distances would appear to become greater and greater as the "scale" of space altered.

It was a nightmare condition, but a very useful one. Nothing could reach a ship carrying the Field; it might be englobed by an enemy fleet yet would be as inaccessible as if it were at the other side of the Universe. Against this, of course, it could not fight back without switching off the Field, but this still left it at a very great advantage, not only in defense but in offense. For a ship fitted with the Field could approach an enemy fleet undetected and suddenly appear in its midst.

This time there seemed to be no flaws in the new weapon. Needless to say, we looked for all the possible objections before we committed ourselves again. Fortunately the equipment was fairly simple and did not require a large operating staff. After much debate, we decided to rush it into production, for we realized that time was running short and the war was going against us. We had now lost almost the whole of our initial gains and enemy forces had made several raids into our own solar system.

We managed to hold off the enemy while the Fleet was re-equipped and the new battle techniques were worked out. To use the Field operationally it was necessary to locate an enemy formation, set a course that would intercept it, and then switch on the generator for the calculated period of time. On releasing the Field again—if the calculations had been accurate—one would be in the enemy's midst and could do great damage during the resulting confusion, retreating by the same route when necessary.

The first trial maneuvers proved satisfactory and the equipment seemed quite reliable. Numerous mock attacks were made and the crews became accustomed to the new technique. I was on one of the test flights and can vividly remember my impressions as the Field was switched on. The ships around us seemed to dwindle as if on the surface of an expanding bubble; in an instant they had vanished completely. So had the stars—but presently we could see that the Galaxy was still visible as a faint band of light around the ship. The virtual radius of our pseudo-space was not really infinite, but some hundred thousand light-years, and so the distance to the farthest stars of our system had not been greatly increased—though the nearest had of course totally disappeared.

These training maneuvers, however, had to be canceled before they were complete owing to a whole flock of minor technical troubles in various pieces of equipment, notably the communications circuits. These were annoying, but not important, though it was thought best to return to Base to clear them up.

At that moment the enemy made what was obviously intended to be a decisive attack against the fortress planet of Iton at the limits of our solar system. The Fleet had to go into battle before repairs could be made.

The enemy must have believed that we had mastered the secret of invisibility—as in a sense we had. Our ships appeared suddenly out of nowhere and inflicted tremendous damage—for a while. And then something quite baffling and inexplicable happened.

I was in command of the flagship *Hircania* when the trouble started. We had been operating as independent units, each against assigned objectives. Our detectors observed an enemy formation at medium range and the navigating officers measured its distance with great accuracy. We set course and switched on the generator.

The Exponential Field was released at the moment when we should have been passing through the center of the enemy group. To our consternation, we emerged into normal space at a distance of many hundred miles—and when we found the enemy, he had already found us. We retreated, and tried again. This time we were so far away from the enemy that he located us first.

Obviously, something was seriously wrong. We broke communicator silence and tried to contact the other ships of the Fleet to see if they had experienced the same trouble. Once again we failed—and this time the failure was beyond all reason, for the communication equipment appeared to be working perfectly. We could only assume, fantastic though it seemed, that the rest of the Fleet had been destroyed.

I do not wish to describe the scenes when the scattered units of the Fleet struggled back to Base. Our casualties had actually been negligible, but the ships were completely demoralized. Almost all had lost touch with each other and had found that their ranging equipment showed inexplicable errors. It was obvious that the Exponential Field was the cause of the troubles, despite the fact that they were apparent only when it was switched off.

The explanation came too late to do us any good, and Norden's final discomfiture was small consolation for the virtual loss of the war. As I have explained, the Field generators produced a radial distortion of space, distances appearing greater and greater as one approached the center of the artificial pseudo-space. When the Field was switched off, conditions returned to normal.

But not quite. It was never possible to restore the initial state *exactly*. Switching the Field on and off was equivalent to an elongation and contraction of the ship carrying the generator; but there was an hysteresis effect, as it were, and the initial condition was never quite reproducible, owing to all the thousands of electrical changes and movements of mass aboard the ship while the Field was on. These asymmetries and distortions were cumulative, and though they seldom amounted to more than a fraction of one per cent, that was quite enough. It meant that the precision ranging equipment and the tuned circuits in the communication apparatus were thrown completely out of adjustment. Any single ship could never detect the change—only when it compared its equipment with that of another machine, or tried to communicate with it, could it tell what had happened.

It is impossible to describe the resultant chaos. Not a single component of one ship could be expected with certainty to work aboard another. The very nuts and bolts were no longer interchangeable, and the supply position became quite impossible. Given time, we might even have overcome these difficulties, but the enemy ships were already attacking in thousands with weapons which now seemed centuries behind those that we had invented. Our magnificent Fleet, crippled by our own science, fought on as best it could until it was overwhelmed and forced to surrender. The ships fitted with the Field were still invulnerable, but as fighting units they were almost helpless. Every time they switched on their generators to escape from enemy attack, the permanent distortion of their equipment increased. In a month, it was all over.

This is the true story of our defeat, which I give without prejudice to my defense before this Court. I make it, as I have said, to counteract the libels that have been circulating against the men who fought under me, and to show where the true blame for our misfortunes lay.

Finally, my request, which as the Court will now realize I make in no frivolous manner and which I hope will therefore be granted.

The Court will be aware that the conditions under which we are housed and the constant surveillance to which we are subjected night and day are somewhat distressing. Yet I am not complaining of this; nor do I complain of the fact that shortage of accommodation has made it necessary to house us in pairs.

But I cannot be held responsible for my future actions if I am compelled any longer to share my cell with Professor Norden, late Chief of the Research Staff of my armed forces.

H. NEARING, JR.

# The Mathematical Vood

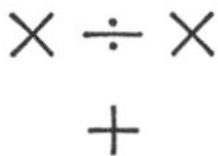

"WHO WAS IT whose slave Socrates extracted the Pythagorean theorem from?" said Professor Cleanth Penn Ransom, of the Mathematics Faculty. "From the mind of, I mean."

"Meno," said Professor Archibald MacTate, of Philosophy.

Ransom's eyes brightened. "That's the name of Plato's dialogue. The one that tells all about it."

"Yes."

"Well, it's a lot of nonsense." Ransom stuck out his little belly and began to swing in his swivel chair.

MacTate smiled. "See here, old boy, you're speaking of the man I teach." He tapped Ransom's desk with his finger. "Do I go about carping at Gauss and Newton—?"

"That's exactly the point." Ransom stopped swinging and aimed a forefinger at his colleague. "You can teach your pupils right. Tell them that wasn't just any old slave Socrates was working on. You can quote me." He jabbed his belly with a thumb.

"Rather an obvious conclusion," said MacTate. "What's the trouble? Has someone flunked algebra?"

Ransom waved disdainfully. "Somebody's always flunking algebra. That's nothing. This man has flunked it six times. Twice in summer school."

"You mean to tell me you could find no substitute for a barbarous requirement like—"

"It wasn't us." Ransom shook his head impatiently. "Some dim-witted dean kept coming across an obsolete catalogue listing that no one ever bothered to take out."

"But the boy—"

"The boy." Ransom groaned. "MacTate, I want you to meet the boy. He's coming here this afternoon. You're a philosopher. Maybe you can figure him out."

"He's in your algebra class now, I take it."

Ransom nodded. "And he's a senior, and it's too late for him to drop a course without getting a failure in it, and he sits—"

"Football player?"

"No, that's the funny thing about it." Ransom looked puzzled. "He's naturally dumb."

There was a decorous knock at the door.

"That's him now. He, I—" Ransom swung his chair around, straightened his tie, laid one hand gracefully on the desk and grasped his lapel with the other. Putting on a grave but benign expression, he intoned, "Come in," with a rising accent.

The young man who stepped into the office wore brown slacks, a blue coat, and a yellow tie. He was slight and narrow-shouldered, but his head and hands were abnormally large. About his eyes was a hunted expression.

"Appointment, doc." His voice was an uncertain baritone.

"Ah, yes. Sit down, Finchell." Ransom waved at a chair. "This is Professor MacTate, of the Philosophy Department."

The young man shook hands with MacTate and sat down.

"Now, Finchell," said Ransom, "what seems to be the—ah—core of your difficulties? With algebra, I mean." He looked at the boy piercingly.

Finchell rubbed his nose. The hunted look about his eyes grew sharper. "Well, I don't know, doc. I'm just a little slow, I guess."

Suddenly an idea blossomed in his eyes. "My father and grandfather were actuaries. Do you think the vein could be worked out? You know, like a mine?"

Ransom looked at him. The hunted look returned to the boy's eyes. He smiled halfheartedly. "No good?"

Ransom reached into his desk drawer and took out a brown book. He flipped the pages. "This, Finchell, is a grade-school mathematics book. On page—twelve—" he turned a page and pressed it down—"we have a problem that most morons should be able to solve. Here. 'If Farmer Brown's horse eats one barrel of oats every two days, what part of the barrel constitutes his daily fare?' In other words, Finchell —that's a little difficult in the phrasing so I'll explain it to you—if this horse eats *one* barrel of oats every *two* days, how much oats does he eat in *one* day?" Ransom noted with joy that the boy's eyes were lit with a dawn of comprehension. "In terms of barrels, I mean," he added. The light in Finchell's eyes died.

"Finchell!" Ransom glared at the boy for a moment, then regained control of himself. He modulated his voice to bell-like tones. "See here." He took a sheet of paper from the desk drawer, drew the outline of a barrel on it, and bisected the outline horizontally. "The horse eats so much of the barrel every two days." He waved his pencil vaguely over the whole outline. "It takes him *two* days to eat so much. See? Now then, in *one* day, he would eat—" he pointed the pencil at the upper half of the outline—"this much. All you have to do is divide one barrel by two days. Right? Now put that down on the paper here. One divided by two." He handed the pencil to the boy.

Finchell looked at the paper as if he had been ordered to jump off a skyscraper. The hunted look about his eyes became poignant.

Ransom smiled at him ingratiatingly. "*One* divided by *two*. You can do it. Go on, write down *one*—"

The boy drew in a sobbing breath and traced a thin vertical line on the paper.

"*That's* it. You've practically got it. Now divide it by two."

Finchell stared at the paper.

"Go ahead. Don't you see? You've practically got it there. How many times does two go into one?"

Finchell dropped the pencil and looked at his mentor with tormented eyes. "It *can't* go in, doc," he said. "It's too damn big."

MacTate hastily pulled out his handkerchief and coughed into it uncontrollably. Ransom stared at his protégé incredulously. Then he dropped the book back into the desk drawer. The boy squirmed in his chair.

MacTate finally controlled his cough and wiped his eyes. "Tell me." He looked at Finchell. "How are you with the multiplication tables?"

Finchell brightened. "Oh, I can do those."

"You can?" Ransom's tone could not quite disguise his skepticism. "Let's see. How much is two— No. Let's make it hard. How much is nine times three?"

The boy fixed his eyes on the ceiling and twisted his jaw off center. For a minute or two he seemed to be chewing an imaginary taffy. Then he spoke. "Twenty-six. No. Twenty-seven."

"My God, that's right." Ransom looked at the boy with wonder in his eyes. "How'd you do that? In your head?"

Finchell dropped his eyes deprecatorily. "Well, yes. Sort of. I did it on my teeth."

"Oh, on your teeth."

"Yes. You see, I figure that one times nine is nine. Everybody knows *that*. So then I put my tongue on this wisdom tooth—" He put a finger into his mouth and pointed. "That's ten. Now I know I have eight teeth on each side of my lower jaw. So the tooth to stop counting with when you're multiplying nine is this one." He pointed. "One past the half of your jaw. You count up to there twice, beginning with ten, and you have three times nine." Finchell smiled with an air of having overcome difficulties reasonably. "It's less noticeable than using your fingers. They don't laugh at you so much."

"Well, Ransom, you can't say there's absolutely nothing to work with there." MacTate turned to his colleague. "He can multiply, and that's a start."

"Yes." Ransom glowered at the boy.

MacTate rubbed his chin. "Perhaps if there were some way of giving him confidence— You know. A simple formula of some sort that he could memorize and apply to various sorts of problems."

Ransom studied his protégé and shook his head judiciously. "A rabbit's foot would work better."

MacTate smiled. "You mean something on the order of a football player's talisman?"

"I've seen it work." Ransom looked at Finchell.

Following his colleague's glance, MacTate noticed that Finchell's eyes were shining with a strange eagerness. He hastened to dispel the boy's unseemly interest in this turn of the conversation. "Now, Ransom. Next you'll be tutoring a wax doll containing his fingernail clippings. Voodoo, or whatever it is."

Ransom turned to him with an expression that matched Finchell's. "What did you say?"

"I merely said that it's absurd to suppose that contagious magic—"

"Wait." Ransom aimed a finger at him. "What's so absurd about it? I've read in—lots of place that that voodoo stuff does funny things sometimes. Who knows?" He looked at Finchell. "Who knows what might be best for this man?" He put a hand on the boy's shoulder. "Anything we can do for him is well warranted." He swung back to the desk. "Short of murder," he added under his breath.

"But Ransom. Don't you think—?"

Ransom gave his colleague a warning glance. "I think it's an idea worth trying." He reached into the desk drawer and took out a fingernail clipper. "Here, Finchell. Let's have some of your fingernails."

Finchell pressed his fists against his stomach and shrank back into his chair.

"What's wrong, man?" said Ransom. "This might work for you."

"I—" Finchell gasped. "I don't have any fingernails, doc. I bit them all off trying to do algebra."

Ransom laughed with strained sympathy. "Is that all? Well, your hair will do just as well." He whipped out a pocketknife, opened it, and sliced off several strands of the boy's hair. "Now. MacTate and I will make this wax doll this afternoon. And tonight—" Ransom clapped the boy smartly on the shoulder—"tonight I wouldn't be surprised if you found mathematical concepts suddenly—generating in your mind." He laughed, rather too heartily. "Tomorrow in class we'll see what's happened."

Finchell got up, clasped Ransom's hand fervently in both of his hands, and looked earnestly into the little man's eyes. "Thanks, doc. Thanks—" He turned and left the room.

When the door had closed behind Finchell, MacTate looked at his colleague. "My dear Ransom—"

"Now, it isn't going to hurt to try this, MacTate. The boy is one of those low mental types that can be helped by superstitions. If we can teach him any mathematics at all, by any means at all—"

"But if he tells anyone, you'll be the laughingstock—"

"*He* won't tell anybody." Ransom waved a deprecatory hand in the direction of the door. "Didn't you hear him say how he counted on his teeth instead of his fingers so people wouldn't laugh at him? He's scared to death people will find out he's dumb."

"But he told you about the teeth-counting."

"All right. That's different. He was confessing to a—diagnostician. But this voodoo—" Ransom made a face.

"Well, I hope you're right. By the way—" MacTate looked at his little colleague curiously—"do you really intend to make a wax doll?"

Ransom looked up with a sneer that slowly faded into an expression of suspicion. "Is there any good reason? Why I should, I mean?"

MacTate looked thoughtful. "Disregarding the ethical consideration, it occurred to me that if you have sized up the boy correctly, he is just the sort to insist on seeing the doll. Better have a few of his hairs sticking out of it, too. If this scheme is to succeed at all, it has to be quite circumstantial."

Ransom sighed. "All right. I'll make a doll." He slapped the desk with his hand. "But I'm *not* going to teach it algebra."

MacTate did not argue the point, but later he wondered if he should not have argued it. From time to time as he lectured to his classes the next day, he would catch his thoughts wandering to Finchell and the wax doll. The day after, he was in Ransom's office again.

"Well, Ransom, how's your boy coming along? What's his name? Finchell." Ransom glared at him.

"No change, I take it." MacTate glanced over the desk. "Do you have the doll here?"

Ransom opened his bottom desk drawer, took out the doll, and set it on his desk. About six inches tall, its body had been painted brown and blue, with a yellow streak to represent a tie. From the top of its head rose a quincunx of hairs embedded in the wax. Its features were vague but somehow sinister.

"You were right about one thing." Ransom turned the doll around to look at its face. "Finchell showed up here the next morning and asked to see this thing. Even wanted to know how I went about teaching it." He smiled reminiscently.

"What did you tell him?"

"Oh, I gave him some kind of double talk. Something about going through the book with it step by step. I don't remember—"

"Did he believe you?" MacTate turned the doll around again.

"What?" Ransom stared at him.

"Was he convinced that you really had tutored the doll?" MacTate waved his hand. "Or do you think he saw through you?"

"Of course he did. Believe me, I mean." Ransom looked confused. "I think he did. How can you tell what a boy like that is thinking? If he thinks at all. He didn't press the point, anyway."

"I see." MacTate brushed his fingers over the hairs on the doll's head.

Ransom's eyes narrowed. "Look, MacTate. What are you getting at? Why shouldn't the boy believe me? I made the doll and showed it to him, didn't I? Isn't that—?"

"Ransom, old boy—" MacTate put his hands on the desk and leaned forward— "As an old friend I may observe without offense that you are one of the world's worst liars. And as I said before, a trick of this sort has to be as circumstantial as possible. Now, the boy's mathematical ineptitude does not necessarily preclude penetration with respect to human reactions. Maybe he can't learn your algebra, but he probably can size *you* up better than you think. If you want my advice, I think you should go through a book with this doll—step by step, as you said—so that you can assure the boy unequivocally—"

"Wait." Ransom looked outraged. "You want *me* to teach algebra to this—this—" He gestured toward the doll. "My God, MacTate. Everybody'll think I'm crazy."

"You pointed out yourself that the boy probably won't mention the arrangement to anyone."

"But, MacTate. Can you picture me, now—"

MacTate shrugged. "Suit yourself, old man. I'm only telling you what I think."

Ransom leaned his elbows on the desk and put his jaws between his hands, looking ruefully into the distance. "All right. I said I would try this fool thing, and I will. Go away, though, MacTate. I won't do it in front of anybody."

Next morning, the ringing of his office phone roused MacTate from a nap induced by the *Journal of Aesthetics*, open on his desk. Ransom's voice, on the other end, was strained and excited.

"MacTate. He worked a problem. This morning."

"Problem? He?" MacTate was not yet fully awake. "Who?"

"Finchell. Who else? He worked a problem in class all by himself. I'm not exaggerating."

"Oh, Finchell. Yes. He worked a problem? What kind of problem was it?"

"An *al*-gebra problem. What's the matter with you, MacTate? You asleep or something?"

"No, no. I meant was it a multiplication problem, as before, or something more difficult. You say it was an algebra problem? Nothing terribly hard, I trust."

"No, nothing terribly hard," Ransom trilled ironically. "Just a little thing involving the binomial theorem, that's all."

"Ransom. You're pulling my leg."

"Look. MacTate. On my honor as a— Look. I swear by everything I—"

"The binomial theorem." MacTate tasted the thought. "You're really serious about this?"

"On my honor as a—"

"And it was entirely correct? No indication that someone had done it for him to memorize—?"

"Absolutely not. I made him do it three times with different signs and once with different exponents. He's got it down. The binomial theorem, I mean."

"Just a moment, Ransom. Did you teach the doll anything yesterday?"

"What if I did?"

"What was it you taught it? Think."

"Why—" Ransom's voice dropped almost to a whisper. "I guess it was the binomial theorem."

A fortnight later Ransom informed his colleague that Finchell, having mastered algebra and analytic geometry and bitten deep into calculus, had transferred to a mathematics major. "We're going to let him satisfy requirements by taking special examinations," said Ransom. "By the end of the year he'll know more mathematics than a lot of members of this department, anyway." He laughed. "What a boy! I still feel like a fool teaching this thing—" he patted the doll on the head—"but it's a—unique experience, covering a book a week and knowing that somebody's learning everything you teach. I almost know how it feels to be a coach."

"Well, you're tutoring a team, so to speak." MacTate smiled.

"And how they click. Maybe we should teach everybody that way."

MacTate shook his head judiciously. "No. Won't do. There are too many geniuses on this campus already."

"But really—" Ransom set the doll on the edge of his desk—"Finchell is a real genius. Or this doll is, I don't know which. Next week I'm going to start them on complex variables." He tripped the doll with his hand and watched it flip over into the wastebasket. "I wonder how long it would take them to learn all the math I can teach." He reached into the wastebasket and set the doll on the edge of his desk again.

MacTate looked at him thoughtfully. "It's possible that mathematical *gestalten* are already forming in Finchell's mind that have never happened to shape up in yours. It's a matter of juxtaposition and attention and general experience, isn't it?"

"But what about Socrates and the man's slave? You remember. In Plato. How you're born with math in your head, and you don't have to learn it but only be reminded of it." Ransom tripped the doll again and sent it spinning into the wastebasket. Its head struck the edge of the metal container with a loud clang.

"Aren't you afraid you'll break that thing, Ransom, playing with it that way? I wonder what would happen if you did."

"It won't break," said Ransom. "Special grade wax. I do this all the time."

"Well, as for Meno's slave—" MacTate's eyes twinkled—"you yourself have assured me that the notion of innate mathematical concepts is untenable. 'A lot of nonsense,' if I remember correctly." He looked at his watch and rose. "So that takes care of that. I have a class in five minutes." He went to the door and then turned around. "Don't forget to take your doll out of the wastebasket. Heaven forfend we should nip a genius in the bud by losing his psychic control."

From time to time during the ensuing months MacTate heard from his rapturous colleague concerning Finchell's new triumphs. Then one day he was summoned by phone to Ransom's office to hear something "terribly important." When he got there, he found the little man smiling with something like transport at a sleek young man in a well-fitting gray suit who sat before his desk. MacTate stared at the young man, trying to place him.

"MacTate, you remember Finchell." Ransom beamed.

As the young man rose to shake hands with him, MacTate almost rubbed his eyes. Gone was the self-consciousness, gone the hunted expression about the eyes, gone the rabbity awkwardness of the mathematical idiot whom he had seen here only a few months before. The person shaking his hand was mature and nearly handsome, radiating intelligence and competence. His handclasp was almost numbing.

"I have not forgotten that my career began with a suggestion of yours, sir. I am happy to see you again." Gone, too, was the uncertain voice. Finchell spoke in an enormous bass.

"Look," said Ransom as the others sat down, "I want some of the credit here, too. I was right that time about Socrates and the slave, MacTate. It's a matter of ability and experience. Mathematical aptitude, I mean. How did you put it? Juxtaposition and attention."

"You mean Finchell knows something you haven't taught him?" MacTate looked at the young man with interest.

Ransom pretended to wince at the understatement. "MacTate, Finchell knows something no other mathematician yet born has discovered. He's solved the Problem of Dirichlet."

"He has? What on earth is that?"

"Dirichlet was Gauss's successor at Goettingen. Among other things, he tried to prove that a region bounded by a single curve, like a slice of the earth's surface, for example, can be projected isogonally and point for point on a circle. In the case of the earth it amounts to reproducing a convex surface on a plane, like a map. Well, to prove it, he tackled an analogous problem in the calculus of variations. The calculus problem was to find a function, u, which with its first derivatives is continuous in the region to be projected, which has continuous second derivatives, and which makes a minimum of the integral—I won't go into details. Anyway, for a while they assumed that a function of this sort really exists, and they called that method of solving the problem Dirichlet's Principle. Then a fellow named Weierstrass showed that the reasoning was insufficient. Now Finchell—" Ransom looked at the young man with almost maternal pride—"Finchell has proved definitively the existence of the function u."

MacTate looked at Finchell and nodded benignly. "Quite something, I imagine."

"The department went over and over it," continued Ransom, "and then sent it every place for checking—Chicago, Princeton, London, every place—and nobody could find anything wrong with it." He beamed at the young man again. "Finchell already has an international reputation."

Finchell laughed, richly and somewhat pompously, and stood up. "Now, Professor, you're likely to make an egotistical ass of me. I'd better get back to my researches before you do so." He seized MacTate's hand, and smiled heartily. "A great pleasure to see you again, sir." He turned briskly and left the room.

MacTate looked after him reflectively. "And just a few months ago—"

"My God, do you remember that?" Ransom screwed up his face. "To think how I was hoping something horrible would happen to him. And now he's the pride of the University. Next week he's going to read a paper on the function u in the public lecture series. Only student they ever let do that. And some of the biggest wigs in this part of the country are coming to hear it."

"Well—" MacTate looked thoughtful—"I'm glad to hear of the happy event. I suppose Finchell's career must lie in mathematics now. I just wonder what he's going to do when you stop teaching that doll. Have you tried to wean him yet?"

"No." Ransom took the voodoo doll out of his drawer, looked at it, and set it on the edge of his desk. "But he won't need it much longer. He's working on a critique of Einstein's unified field theory—you know, about gravitation and electromagnetism being the same thing. Going to read a paper on it at the convention next summer. So I've got to take him through complex tensors. And then we can pull this hair out and—" he flipped the doll into the wastebasket—"throw this away."

"Have you mentioned that to him?"

"No. Why should I? He's doing all right just the way things are."

"You don't think he might resent your proposal?"

Ransom took the doll out of the wastebasket. "I don't see why he ever has to know about it. He hasn't asked about the doll for a long time now. Probably forgotten about it. As a matter of fact—" his eyes twinkled—"Finchell seems to be interested in a different kind of a doll lately. Girl that sings downtown. Name of—Dolores something. Anyway, he's worked up such an interest in music that I'm almost jealous." He grinned.

MacTate waved at the doll. "I wonder if that's jealous, too."

"MacTate. Will you stop worrying about the doll. Anybody would think you took this voodoo thing seriously. I have to keep fooling with it so he won't think I'm lying to him, but that's no reason to carry on as if there were something—valid in it. I give him the same assignments I teach the doll, and he works them out for himself, that's all."

Ransom set the doll on the edge of the desk. "Someday when he's a doddering old professor he might remember and say, 'Ransom, my old friend and benefactor, what ever happened to that silly wax voodoo you made of me?' And I'll clap him on the shoulder and say, 'Finchell, you were dreaming. There never was any such thing. All you needed was a little confidence, and—' " he flipped the doll into the wastebasket—" 'I gave it to you.' So stop worrying."

MacTate wished that Ransom's blandness were contagious. He could not overcome a sense of foreboding, a feeling that the whole thing had been wrong to begin with and was now out of hand. But he consoled himself with the reflection that it was not really his affair, and for the next week he avoided his little colleague's office so that he would not have to think about the matter.

But then one morning his office phone woke him from a *Journal of Aesthetics* doze again, and Ransom was wildly insisting that he come over at once.

"What is it this time?" MacTate said sleepily. "Has Finchell discovered the thirteenth dimension?"

"MacTate. Weren't you at the lecture last night? I know I told you—"

"What lecture? You mean Finchell's? On the function—what was it? No, I'm afraid I wasn't there. I—"

"Well, neither was Finchell."

"What?"

Ransom had hung up. MacTate lost no time in getting over to his office. The little man was pacing restlessly up and down.

"MacTate, why should he do this to me? Why? I make a great mathematician of him, have his problem checked for him, put him in the lecture series and invite the bigwigs to hear him. And then he disappears. Without a word."

"What did you do? Cancel the lecture?"

"Couldn't. Everybody was already there. We had to let the Dean talk about methods of teaching rapid calculation. It was—dismal." Ransom sat down and held his head in his hands. "Mathematically speaking, the University is in the doghouse."

MacTate looked thoughtful. "When did you last see Finchell?"

"Let's see. This is Tuesday." Ransom paused a moment. "Yesterday I figured he was resting up for the lecture, so I didn't look for him.

The weekend doesn't count. Thursday and Friday I was out of town. I guess it's been almost a week."

"Old man—" MacTate put a hand on his colleague's shoulder— "have you inquired at the jails? Or the—hospitals?"

Ransom gasped. He looked past MacTate with glazed eyes. His lips formed the word "morgue." He grabbed his hat and darted to the door. "Let's get down there."

After a week or so of frenzied inquiry, between and after classes, at the morgue, the bureau of missing persons, the police department, the public health service, and five or six insurance companies, Ransom was beginning to suspect that Finchell had been shanghaied for service on a ship engaged in illicit trade, while MacTate favored the theory that the doll had contracted contagious amnesia from striking its head on the rim of the wastebasket.

"You know," he said one day as they sat in Ransom's office, wearied by the usual rounds, "I wonder if there *could* be any connection between the doll and Finchell's disappearance. If he sensed that it wasn't cared for properly—"

"Who doesn't care for it properly?"

"When you flip it into the wastebasket, you know, it sometimes strikes its head against the edge. Have you ever noticed whether it's chipped or—?"

"Of course it's not chipped." Ransom looked offended. "I should know, shouldn't I, working with it all the time?"

"When was the last time you did work with it?"

"Why, it was— What does that matter? Look, I'll show you." Ransom opened the bottom desk drawer and reached into it. "You can see for yourself that it's just the same—" He opened the drawer wider, bent over it, and rummaged about in it. "Funny, I'm sure— Maybe I put it in this one." He opened the drawer above and rummaged in it. Then the drawer above that, and finally the central drawer at the top of the desk. He looked at MacTate in bewilderment. "What happened to it?"

"How about the wastebasket?"

They both leaned over the wastebasket, nearly bumping heads. Ransom reached in and threw out several balls of paper and a candy wrapper. There was no doll.

"Well," said MacTate, "there may be more to this than—"

"MacTate. Listen. What could have happened to that doll? We've got to find it. We've got to find the person that stole it." Ransom's eyes were anguished. "But who would want to steal it?" He wrung his hands.

There was a knock at the door.

"Now who—?"

The door opened slightly and a head appeared around it.

"Professor Ransom?" The visitor came into the room. He wore blue slacks and a maroon jacket, and his silk shirt was open at the neck. In spite of the waving hair and newly grown mustache, Ransom recognized his erstwhile protégé.

Finchell moved languorously to a chair, dropped into it, and smiled fatuously at Ransom. "I'm leaving the University, Professor, and remembering that you were my adviser, I felt that I ought to let you know." He spoke with a peculiarly meticulous articulation and resonant, pear-shaped tones.

"Well now, that's damned decent of you, Finchell." Ransom was unable to sustain his sarcasm. "Where have you been, you Judas?" he burst out. "Why did you disgrace me at the lecture? Why did you keep me running to the morgue and the—?" As if suddenly realizing the futility of his rage, he stopped and looked at Finchell appealingly. "Finchell—why?" he whispered.

Finchell looked mildly astonished at Ransom's outburst. "Lecture? At the morgue, you say?" He squeezed his eyes shut and drew a hand gracefully across them. "Yes. I remember. There was something about a lecture. But not at the morgue, was it? Well—" he opened his eyes—"I trust I missed nothing indispensable."

Ransom was speechless. MacTate took over. "You say you're leaving the University?"

"Yes." Finchell put on a supercilious expression. "Not that I disapprove of the sort of work you people do here. It has its place. But as Dolores says, when one's art is at stake—" He smiled tolerantly. "I'm to have my final audition tomorrow, and waiting for a degree from the University would delay my career for some months. Not that I disapprove of degrees, as I said, but—" He gestured gracefully with his hand. "You see how it is." He stood up. "Nice knowing you, Professor," he said to Ransom. "I'll try to remember to send you tickets sometime." He turned to the door.

"Finchell." MacTate called after him. One last thing before you go. Does the function *u* mean anything to you?"

Finchell turned around. "The function *u*?" He squeezed his eyes shut and touched them gracefully with his hand. "Afraid not." He opened his eyes. "Sounds like one of those frightful mathematical things. I never could do math." He turned again and swung from the room.

MacTate sighed. "Well, I hate to say I told you so, but I knew you should have been more careful of that doll."

Ransom started: "That doll! MacTate, call him back. We forgot to ask him what he did with the doll."

"What makes you think he did anything with it?"

"But we're the only ones who knew. Who else but—"

"I'm not so sure." MacTate shook his head. "Anyway, my guess is that the person who empties your wastebasket is the one who can tell us most about the doll. Who do you suppose that would be?"

Ransom looked at him. "Do you think—?" He stood up. "Let's find out."

From the Director of Maintenance, they went to the Superintendent of Buildings and Grounds, the Supplies Coordinator, the Foreman Janitor, and the Assistant in Charge of Washrooms and Waste—and finally found the emptier of Ransom's wastebasket filling soap containers in the School of Business Administration. He was a wiry man of indeterminate age.

"Doll?" he said in answer to their questions. He rubbed his nose reflectively with his forefinger. "Oh, the little doll. Painted with colors. Yes. I find him in wastebasket. Two, maybe three weeks ago sometime. I am taking him home for the baby, but he's lost." He shook his head sadly.

"How could you lose a thing like that?" Ransom was annoyed. "It was at least—"

"Just a moment, old man." MacTate stepped forward. "Tell me," he said to the janitor, "did you lose the doll at home?"

The janitor pondered, then shook his head. "No. When I get home, he's gone. I don't tell the baby. She's—"

"Where did you go after you left the University with the doll?"

The janitor turned his head, pursed his lips, and pressed a finger against them. Suddenly his eyes lit up. "Ah. Now I remember. I go to the opera, where it plays *Meistersinger* by Richard Wagner. In the second act I go. I am apprentice."

"You mean where all the apprentices come out and riot because what's-his-name is courting one of their fiancées by mistake?"

The janitor smiled. "That." He nodded. "That is right."

"And did you have the doll with you when you went on the stage?"

"Yes. I have him in my pants pocket." He patted the seat of his overalls. "Wrap up in—I think brown paper. Under my—what? Costume."

"Then after you came off the stage, did you look to see if the doll was still there?"

The janitor shook his head and held a fist to one eye. "When I am come off the stage, I have the—what? Black eye. They forget it is only play. I don't think about nothing else till I am home. Then he's gone. The doll."

"But you think you must have lost it during the riot scene? On the stage."

"I go back after and look. Next day. I am smuggler in *Carmen* by Georges Bizet. I ask stagehands." The janitor shrugged. "They do not see. They think, I think too, some star pick him up. Stars very—what? Superstitious. They find something on stage, anywhere, they pick him up. Hide for good luck. Never tell nobody."

"Well, Ransom." MacTate turned to his colleague. "That's that. Your voodoo is probably sitting in some diva's dressing room, listening to arias and scales—" He stopped, suddenly struck by a thought. "Just a moment. Didn't you tell me Finchell has a friend who sings? What kind of singer is she? Do you know?"

Ransom frowned impatiently. "What's that got to do with—?" His mouth fell open. He looked at MacTate, then jabbed a finger at him. "Opera."

"Wouldn't it be a coincidence—?" MacTate shrugged.

Ransom groaned. "An opera singer. Oh, God. And to think he was ready to tackle Einstein. MacTate, we've got to get that doll back." His eyes blazed. "I'll get a search warrant—"

MacTate shook his head. "No use, old man. They'd only make a fool of you." He looked thoughtful. "Anyway, didn't Finchell leave notes of any sort on this electromagnetism thing? Illustrative figures or something like that?"

Ransom nodded ruefully. "I've got his notes," he said, "and they look like the greatest thing since the Theory of Special Relativity. But nobody will ever know now."

"Why not?"

"Well, to save time in his figuring, Finchell invented two new symbols. Without bothering to put down what they mean. Crazy things. One he called a '*horse*.' "

MacTate looked at him, startled. "And the other—?"

"A '*barrel*.' What I don't get—" Ransom frowned with perplexity —"is where he could have picked up crazy names like that."

FREDRIC BROWN

# Expedition

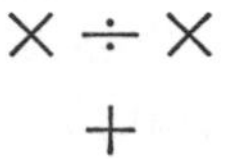

"THE FIRST MAJOR EXPEDITION to Mars," said the history professor, "the one which followed the preliminary exploration by one-man scout ships and aimed to establish a permanent colony, led to a great number of problems. One of the most perplexing of which was: How many men and how many women should comprise the expedition's personnel of thirty?

"There were three schools of thought on the subject.

"One was that the ship should be comprised of fifteen men and fifteen women, many of whom would no doubt find one another suitable mates and get the colony off to a fast start.

"The second was that the ship should take twenty-five men and five women—ones who were willing to sign a waiver on monogamous inclinations—on the grounds that five women could easily keep twenty-five men sexually happy and twenty-five men could keep five women even happier.

"The third school of thought was that the expedition should contain thirty men, on the grounds that under those circumstances the men would be able to concentrate on the work at hand much better. And it was argued that since a second ship would follow in approximately a year and could contain mostly women, it would be no hardship for the men to endure celibacy that long. Especially since they were used to it; the two Space Cadet schools, one for men and one for women, rigidly segregated the sexes.

"The Director of Space Travel settled this argument by a simple expedient. He— Yes, Miss Ambrose?" A girl in the class had raised her hand.

# Expedition

"Professor, was that expedition the one headed by Captain Maxon? The one they called Mighty Maxon? Could you tell us how he came to have that nickname?"

"I'm coming to that, Miss Ambrose. In lower schools you have been told the story of the expedition, but not the *entire* story; you are now old enough to hear it.

"The Director of Space Travel settled the argument, cut the Gordian knot, by announcing that the personnel of the expedition would be chosen by lot, regardless of sex, from the graduating classes of the two space academies. There is little doubt that he personally favored twenty-five men to five women—because the men's school had approximately five hundred in the graduating class and the women's school had approximately one hundred. By the law of averages the ratio of winners should have been five men to one woman.

"However, the law of averages does not always work out on any one particular series. And it so happened that, on this particular drawing, *twenty-nine* women drew winning chances, and only *one* man won.

"There were loud protests from almost exeryone except the winners, but the Director stuck to his guns; the drawing had been honest and he refused to change the status of any of the winners. His only concession to appease male egos was to appoint Maxon, the one man, captain. The ship took off and had a successful voyage.

"And when the second expedition landed, they found the population doubled. Exactly doubled—every woman member of the expedition had a child, and one of them had twins, making a total of exactly thirty infants.

"Yes, Miss Ambrose, I see your hand, but please let me finish. No, there is nothing spectacular about what I have thus far told you. Although many people would think loose morals were involved, it is no great feat for one man, given time, to impregnate twenty-nine women.

"What gave Captain Maxon his nickname is the fact that work on the second ship went much faster than scheduled and the second expedition did not arrive one year later, but only nine months and two days later.

"Does that answer your question, Miss Ambrose?"

MILES J. BREUER, M.D.

—

# The Captured Cross-Section

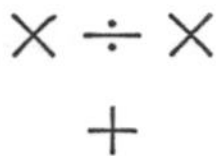

THE HEAD of Jiles Heagey, Instructor in Mathematics, was bent low over the sheets of figures; and becomingly close to it, leaned the curly-haired one of his fiancée, Sheila Mathers, daughter of the Head of the Mathematics Department. Sheila was no mean mathematician herself, and had published some original papers.

"Are you trying to tell me that this stuff makes any sense?" she laughed, shaking her head over the stack of papers.

"Your father couldn't follow it either," Heagey answered. "He used abusive language at me when I showed it to him."

"Now don't be mean to my father. Someday you'll learn that under his blustering exterior he has a heart of gold. But what do these things mean, and what did you bring me in here for?"

"You have followed through Einstein's equation for the transformation of coordinates, have you not?" Heagey explained. "Well, this is Einstein's stuff, only I've carried it farther than he did."

"It doesn't look the same—" Sheila shook her head.

"That is because I am using four coordinates. The most complicated existing equations, with the three coordinates x, y, and z, and involving three equations each with the variables:

$$\begin{matrix} x_1, & y_1, & z_1, \\ x_2, & y_2, & z_2, \\ x_3, & y_3, & z_3, \end{matrix}$$

require that you keep in mind nine equations at a time. That is a heavy burden and relatively few men are able to do it. Here I have four coordinates, w, x, y, and z, and the variables:

$$\begin{matrix} w_1, & x_1, & y_1, & z_1, \\ w_2, & x_2, & y_2, & z_2, \\ w_3, & x_3, & y_3, & z_3, \\ w_4, & x_4, & y_4, & z_4, \end{matrix}$$

requiring that I carry in my mind sixteen equations at one time. That may seem impossible, but I've drilled myself at it for two years, and gradually I was able to go farther and farther—"

"But there are other quantities here," Sheila interrupted, studying the paper intently, "that do not belong in equations for the rotation of coordinates. They look like the integrals in electromagnetic equations."

"Good for you!" Heagey cried enthusiastically. "That pretty little head has something on the inside, too. That is just exactly what they are: electromagnetic integrals. You see, the rotation of coordinates looks very pretty in theory, but when you hook it up with a little practical dynamics—don't you understand yet?"

Sheila stared at the young mathematician in questioning wonder.

"Sheila, jewel, you're just irresistible that way. I can't help it." He gathered her in his arms and kissed her face in a dozen places. She pushed him away.

"No more until you tell me what this is about. I mean it!" She stamped her foot, but a merry smile contradicted her stern frown.

"You're just like your father when you're like that," he said, taking up the papers again. "Very simple little conception," he continued. "Why be satisfied with rotating coordinates on paper? Here's a way to rotate them in concrete, physical reality.

"Listen now. When you rotate two coordinates through ninety degrees, you have an ordinate where there previously was an abscissa. If you rotate three coordinates through ninety degrees, you can make a vertical plane occupy a horizontal position. Now—suppose you rotate four coordinates through forty-five degrees: you can then make a portion of space occupy a new position, outside of what we know as space. And we can bring into this space of ours a portion of the unknown space along the fourth coordinate—"

"The fourth dimension!" gasped Sheila.

"There it is on paper. But we're going to do it in reality. There—" pointing across the room—"are the coils by means of which we can rotate some real space. I want you to see the preliminary trial. As I do not know just exactly what may happen, I am going to rotate only a small portion to begin with."

Sheila's eyes gleamed with excited comprehension.

"Call father in. He's just across the corridor—"

"Not for the very first trial. I want you to see that alone. After we know what it will do—"

"But it may be dangerous. Something may happen!"

"You think it might injure the furniture or damage the building? For the preliminary trial I shall rotate it only for an instant and turn it back instantly."

She clung to his arm nervously while he grasped the black handle of the switch and threw it down, waited a few seconds, and pulled it out again.

They saw nothing. There was a crash, instantaneously loud, and fading almost instantly to a distant, muffled rumble, and ceasing suddenly. There was a heavy thud and a pounding on the floor. Sheila gave a little scream.

There in front of them was a rapidly moving object; it bounced up and down off the floor to a height of three feet about once a second. It did not have the harmonic motion of a bouncing body, however; it stopped abruptly up in the air and shot downward at high speed, hit the floor, stopped a moment and shot back upward. Then it stopped suddenly and hung in the air. It was about the size of a large watermelon, and looked for all the world like human skin: smooth, uniform, unbroken all around.

The two stared at it amazed. Heagey walked up and touched it with the tip of a finger. It grew smaller. And suddenly it decreased to about one-half its former size, retaining its surface smoothness and uniformity unchanged.

It had felt soft and warm, like human flesh.

Now it was increasing in size again, while they stared gasping, speechless, at it. When it stopped growing suddenly, it was the size of a big barrel, with rounded ends. There was a bulging ridge around the middle, on each side of which was a dark brown strap of something like leather. The rest of it was just naked skin.

Sheila and Heagey stood rooted to the spot, staring at it and at each other. What was the thing? Where had it come from?

The Thing began thumping up and down off the floor again, with great, thudding shocks. After a while it desisted and lay still. It was a most uncouth, hideous-looking thing: a great lump of naked flesh with two straps around it. It looked exactly like some huge tumor in a medical museum, or like some monstrosity of birth. Could it be alive?

Both of them approached it cautiously. Heagey pricked it with a pin. The skin was tough and he jabbed hard. A drop of blood appeared.

Then there was a terrible commotion. The object decreased in size to a small sphere like a baseball. In fact, there were several baseball-sized lumps of flesh all around; just naked flesh. They moved rapidly, and two of them were between him and Sheila. Two or three were on the far side of her. He counted ten of them altogether. Five of them closed swiftly around her. Then she was gone!

Her scream, cut suddenly short, still rang in his ears. And she was gone! Suddenly vanished from in front of him! He groped about, feeling for her in the empty air, but there was nothing anywhere. There lay the watermelonlike lump of flesh that he had first seen. It was on the floor and lay quite still. And she was gone! He held his head distractedly.

The door opened and Professor Mathers, Sheila's father, came in.

"What's going on here?" he demanded, blinking his eyes.

Heagey stared blankly, trying to think.

"This thumping and screaming?" the professor continued.

"I think I begin to understand," Heagey began.

"Think you understand!" the professor shouted. "What have you done to my daughter? She doesn't scream for nothing."

He caught sight of the ovoid lump of flesh. He turned pale and stopped as if frozen. Some terrible thought crossed his mind, connecting it with his daughter; had some nefarious experiment turned her into that thing?

"What's that?" he snapped savagely.

"Something's got to be done," Heagey said, chiefly to himself. "We've got to bring her back here. I'm afraid to manipulate the thing too many times; the Lord only knows what else it may dip up."

The professor glared.

"You sound like a first-rate manic-depressive crazy man—"

"Wait till I shut that thing up," Heagey said, getting a hold on himself, "and I'll explain all I know about this. I was getting ready to try to rotate a dog out of space, and so I have a new, strong dog-cage here."

He set the dog-cage down beside the lump of flesh; very gently, very slowly, he pushed it in. His touch recoiled at the warm, soft feel of it; but he got it into the cage and locked the door. Then he set out a chair for the professor, but his hand shook, for his mind was on Sheila.

He sat down facing the professor, his back to the cage. Suddenly the professor's face fell, and his eyes stared ahead with a look of utter blankness. Heagey whirled around and looked at his "specimen." It was out of the cage!

There hadn't been a sound. His eyes had not been off it for ten seconds. The cage was still locked. There it lay, three feet away from the cage, only it wasn't the same. There were two pieces of it now, long, cylindrical, rounded at the ends. Like a couple of legs without knees or feet. Heagey got up and unlocked the cage, noting that it required fifteen seconds. He felt around inside the cage with his hands, but found nothing.

"After all," he sighed, "it is very simple."

The professor stared at him, now thoroughly convinced that he was crazy. Heagey explained about his sixteen equations and how readily they interlocked with the electromagnetic integrals, and of how the very simple application of any form of electromagnetic energy would rotate four coordinates.

"I wanted her to see the preliminary experiment. I used but little power on a small field. Just opened a little trap door into space, so to speak. There is only one explanation for what has happened here. I rotated a portion of a fourth dimension, and left a hole in hyperspace for an instant. Just as if you rotate up a portion of this floor, there will be a hole left. As chance would have it, just at that moment some inhabitant of hyperspace came along and stumbled into it, and I swung back on him and caught him.

"Here he is, stuck. What we see and feel is a cross-section of him, a solid cross-section of that part of him that is cut by our three-dimensional space. See! If I stick my finger through this sheet of paper, the two-dimensional inhabitants on its surface will perceive only a circle.

At first the nail occupies a portion of its circumference; as I push my finger on through, the nail is gone, and folds and ridges appear and disappear. If my whole hand goes through, the circle increases greatly in size. If they draw a circle around my finger and try to imprison it, I can withdraw it and stick it through somewhere else, and they cannot understand how it was done—"

"But what about Sheila? Where is she?"

Heagey's face dropped. He had been full of interest and exultation in his problem. The reminder of her was an icy shock.

"There is only one possible conclusion," he went on in a dead voice. "The struggles of the fourth-dimensional creature swept her out into hyperspace."

The professor sprang up and walked rapidly out of the room. There was something determined in his stride. He slammed the door. Heagey sat down and thought. Somehow he must rescue Sheila.

How could it be done? Should he try the rotation again? He had all the figures and could repeat it accurately. But that would not be at all certain to get her back. The captured fourth-dimensional creature might get away. Heagey didn't want to lose him. Not only that he wanted to study him, but somehow he felt that he must hang on to the only link with that world where Sheila was now lost.

The thought of its getting away worried him. How could he make sure that it would not escape? He reasoned back to the plane section of a three-dimensional object. Enclosing it in a circle would do no good. But, if tied tightly with a circle of rope, it might be kept from moving up and down. Analogically, if he could get this thing into some sort of a tight bag, he might feel free to flip his trap door once more. Ah! then came the brilliant idea!

He could sally out into hyperspace and look for Sheila!

He got the lump of flesh fastened up tight in a canvas sack and lashed the other end of the stout rope with which he tied it around a concrete pillar. Then the door opened and two policemen walked in, followed by the professor. He was urging them on. "There he is! Grab him!" he seemed to say in attitude and gesture, though not in words.

A pang of alarm shot through Heagey. He was needed right here to rescue Sheila. What would become of her if they locked him up? His mind, as usual, worked quickly and logically, in contradistinction to the professor's, who seemed to have been thrown into an unreasoning rage by his daughter's disappearance. He sprang to his switchboard and shouted: "Stop!"

Something in his determined attitude alarmed the policemen; his hand on the ominous-looking apparatus might mean something. They stopped.

"What's this? What do you want?" Heagey demanded.

The professor's torrents broke loose.

"He murdered my daughter. Made away with her. I've got a warrant for his arrest. Nonsensical twaddle about the fourth dimension. Prosecute him to the limit: that's what I'll do. Been hanging around her too much. He's crazy. Throw him in jail. Make him bring her back!"

Heagey laughed a desperate laugh, which made the other three more certain that he was a dangerous maniac.

"Like throwing debtors into jail," Heagey derided acidly. "Fat chance of paying the debt then! Move another step and I'll throw the three of you into unknown hyperspace."

They were all afraid, of they knew not what. Heagey outlined to them that he wanted to go out into hyperspace and search for Sheila. But he would tie himself on a rope fastened at this end. And he wanted someone here at this end, who was friendly to him, to manage things. He telephoned out for a rope and for two of his students. The policemen watched, too puzzled to know what to do. The professor acquiesced, more from fear, like a man at the point of a gun, than because he saw the reason of it.

The rope was delivered and the two students, Adkins and Beemer, arrived. They helped him fix a firm sling around his shoulders, waist, and thighs. The loose rope was coiled up on the floor, several hundred feet of it, and the other end tied to a concrete pillar. There was some amazed staring by the students at the writhing thing in the canvas sack.

"I'll tell you about that later," Heagey said. "All the pointers and dials are set. All you need to do is to throw this switch and jerk it back at once. Adkins, you do that; and, Beemer, you watch the rope. When I signal by jerking it six times, Adkins, you throw the switch again the same way."

That was all. Without another word Adkins threw the switch.

There was the same crash, instantaneously muffled and almost suddenly fading away as at a distance. There was a momentary sensation of agitation, though nothing really moved.

Heagey was gone. The loose end of the rope that had tied him lay on the floor. It was certainly a breathless thing. The professor stared with a sort of vacant expression on his face, as though the solid ground had suddenly dropped from beneath his feet. It dawned upon him that perhaps Sheila had really disappeared that way.

Beemer picked up the end of the rope. It was not an end; it merely looked that way. There was a strong tension on it; in fact it soon began to slip through his hands, and coil after coil was drawn off the pile on the floor and simply vanished. For a while it stopped and then went on unwinding.

The policemen gazed blankly. They were unable to understand what had happened. The man they were to arrest had suddenly melted from sight. They mumbled astonished monosyllables to each other. But they were not as astonished as was Professor Mathers. They did not grasp the enormity of what was going on, as he did. It upset his whole mental universe. He sat awhile and then paced nervously up and down the vast room. He came and looked at the rope. Then he looked at the canvas sack. The sack lay loose as though the contents had escaped. He felt of it and found that it contained three soft baseball-sized objects. He jumped back and shrank away from it. The time seemed interminable. He waited and waited.

Besides an occasional mumble between the policemen or a short exclamation from Adkins or Beemer, there was no conversation. Beemer watched the rope closely. There was a tense nervous strain created largely by the professor's distracted movements. Then, after what seemed hours, though in reality less than one hour, there were six short tugs on the rope. Adkins threw his switch, and out of the crash and tremor Heagey tumbled out on the floor, all tangled up in coils of rope.

He was breathless, haggard, wild-eyed. He lay for a moment on the floor, panting. Then he sprang up and gazed fiercely, wildly about. He seemed suddenly to perceive where he was. An expression of relief came over his face; he sighed deeply and sank down to a sitting position. He looked exhausted; his clothes were disarranged and ripped in some places, and were covered with dust.

The five people looked at him in silent amazement. He looked from one to the other of them; it was a long time before he spoke.

"Good to be back here. I can hardly believe I'm really back. Never again for me."

"What about Sheila? Where is she?" the professor demanded.

Heagey recoiled as though from some shock. He sank again into profound depression. At first he had seemed a little happy to get back. Apparently Sheila had been forcibly driven out of his mind for the time.

"Let me tell you about it," he began slowly. He seemed not to know just how to proceed. "That is, if I can. I don't even know how to tell it. I know what it must feel like to go insane.

"I heard the switch go down as I gave Adkins the signal. Then it seemed like an elevator starting, and that was all. Until I looked around.

"I was sitting on something that looked like rock or cement. Not far from me was that barrellike lump of flesh with the two straps around it, just exactly as I had seen it in the laboratory. And then a row of shapes reaching into the dim, blue distance. The nearer ones seemed to be of concrete or cement. You've heard me jeer at the crazy, cubistic and futuristic designs on book wrappers and wallpaper. Well, those are pleasant and harmonious compared with the dizzy, jagged angles, the irregular, zigzag shapes with peaks and slants, and everything out of sense and reason except perspective. Perspective was still correct. Just a long, straight row fading into the distance. What in the world it could be, I hadn't the faintest idea. However, I gradually reasoned it out.

"Naturally, since I am a three-dimensional organism, I can only perceive three dimensions. Even out in hyperspace I can only see three dimensions. What I saw must therefore be the spatial cross-section of some sort of buildings. I couldn't see the entire buildings, but merely the cross-section cut by the particular set of coordinates in which I was. Now it occurs to me, that since that barrellike thing looked exactly the same to me out there as it did in the room here, I must have been in a 'space' or set of coordinates parallel to the ones we are in now.

"Imagine a two-dimensional being, whose life had been confined to a sheet of paper and who could only perceive in two dimensions, suddenly turned loose in a room. He could only see one plane at a time. Everything he saw would be cross-sections of things as we know them. Wouldn't he go crazy? I nearly did.

"I first started out to walk along beside the row of rocklike shapes. Suddenly near me there appeared two spheres of flesh, just like this one we have here. They rapidly increased in size, coalesced into a barrel-shaped thing with a metal-web belt around the middle, and then dwindled quickly; there were three or four smaller gobs of stuff and then ten or a dozen little ones; finally an irregular, blotchy, melon-like thing which quickly disappeared. In fifteen seconds it had all materialized and gone.

"I was beginning to understand the stuff now. Merely some inhabitant or creature of hyperspace going by. As he passed through my particular spatial plane, I saw successive cross-sections of him. Just as though my body were passing through a plane, say feet first: first there would be two irregular circles; then a larger oval, the trunk, with two circles, the arms, at the sides and separate from it; and so on until the top of the head vanished as a small spot.

"I followed down the line of buildings, looking around. Bizarre shapes appeared around me, changing size and shape in the wildest, dizziest, most uncouth ways, splitting into a dozen pieces and coming together into large, irregular chunks. Some seemed to be metal or concrete, some human flesh, naked or clothed. In a few minutes my mind became accustomed to interpreting this passage of fourth-dimensional things through my 'plane' and I studied them with interest. Then I slipped and fell down. Down I whizzed for a while, and everything about me disappeared.

"I found myself rolling; and sitting up, I looked around again. There was nothing. I still seemed to be on cement or stone; and in all directions it stretched away endlessly into the distance. It was the most disconcerting thing I had ever seen in my life. I was just a speck in a universe of cement pavement. I began to get panicky, but controlled myself and started to walk, feeling the reassuring pull of the rope behind me. I walked nervously and saw nothing anywhere. Evidently I had slipped off my former 'plane' and gotten into a new one. The rope tightened suddenly; perhaps I had reached the end of it. It jerked me backwards and I swung dizzily, my feet hanging loose.

"I swung among millions of small spherical bodies disposed irregularly in all directions about me, even below. They moved gently back and forth in small arcs; and there were large brown bodies—

"Why go through it all? I stumbled from one spatial plane into another. Each seemed a totally different universe. I couldn't get them correlated in my mind into any kind of a consistent whole at all. For a long time I climbed over some huge metal framework; I ran into moving things that grew larger and disappeared; I struggled through a jungle of some soft, green, vegetable stuff. Just all of a sudden I made up my mind that I'd never find Sheila.

"She might be within a foot of me all the time, yet I couldn't get to her, because I couldn't see out of three dimensions. I yelled her name until I was hoarse and my head throbbed, but nothing happened. I grew panicky and decided I wanted to go back. I pulled on the rope and dragged myself toward the direction from which it came; sometimes I slid rapidly toward it; at others I could feel myself dragging my entire weight with my arms. Then I could go no further, pull as I might. It seemed like trying to reach an inch higher than you really can; I couldn't quite stretch that far. So I gave it six short tugs. Very quietly I tumbled out here. I haven't seen Sheila."

The professor was calm. His face was set hard.

"Either you're telling the truth or you're insane as a loon," he said, and his voice was puzzled and sincere. "Perhaps I'm crazy, too. I'm broad-minded enough to admit that is possible. I've got you charged with murder. But I'll give you a chance. What are you going to do about Sheila?"

Heagey's eyes blazed.

"You can go to hell with your chance," he roared. "I want Sheila back worse than you do. If anyone can get her back, it is myself. If you interfere, you simply guarantee that she's lost, that's all. If you want to see her again, keep your hands off! See?"

The professor was a better man than his blustering actions might lead one to think.

"Well, I'm worried," he said shortly. "Can I help you any?"

Heagey never changed expression.

"Perhaps you can. I may need more money than I've got. Just now you can help me most by getting out of here and taking everybody with you and letting me think. I've got an idea. I'll phone you when I want something."

"Well, remember you're charged with murder, and there will be a police guard around this place."

How great and yet how small men will be under trying conditions!

Heagey, left alone, sat and thought. He jumped up and ran his hands through his hair.

"God! Think of it!" he gasped. "Sheila out there alone! In that mad place! Not even a rope!"

He paced rapidly around the room. Then he seized paper and pencil and began to draw. He drew circles and ellipsoids of different sizes and laid the drawings in a row. The professor came in an hour later and found him at it.

"How do you ever expect to find her that way?" he growled peevishly.

"Shut up!" Heagey snapped, his nerves tautened into disrespect. He swept up the papers with his hand and crumpled them into the wastebasket. "No use. Can't study four-dimensional stuff on a two-dimensional plane. Say!" he shouted roughly at the professor, "get me a hundred pounds of modeling clay up here. Quick as you can!"

The professor trotted out after it without a word, much less with any understanding of what it was about.

"Do you think you'll do it?" was his eager attitude one moment, and "If you don't, you go on trial for murder," he raved a moment later.

Far into the night Heagey worked with modeling clay, molding the forms that had appeared in the laboratory and some of those he had seen in hyperspace. He tried to recollect the order in which the various shapes had appeared to him, and laid them in rows in that order. Late into the night he modeled and arranged and stared and studied. Near midnight the professor poked his head in the door.

"She's really gone," he moaned. "She hasn't come home. She's nowhere!" He turned on the haggard Heagey. "The policemen are on the job, so don't try to get away. But I'm offering five thousand dollars to anyone who brings Sheila back."

Heagey snatched a few hours' sleep on the floor. In the morning when the professor opened the door, he was arranging clay balls and clubs into rows and staring at them. As soon as the professor's head appeared, he shouted:

"I've got it! The biggest photographs you can get of Sheila. Head and full-length both. And fast! Hurry!"

He now turned his attention to the object in the canvas sack. He untied the rope from the fourteen-ounce duck, tied the corners of the canvas together, inserted a stout stick (obtained by breaking the leg off a chair), and twisted it, squeezing the small ball of flesh unmercifully. At first sight it was a cruel-looking procedure, but there was method in it. The Thing began to jump back and forth excitedly. He loosened the bulk of his pressure, but kept up a steady, firm tension. His strength was sufficient to hold it fairly steady. Suddenly he loosened all pressure. The mass of flesh suddenly grew larger and the satisfied expression in Heagey's face showed that was what he was working for. Just as when you push hard against someone and then suddenly let go: he falls toward you.

He persisted steadily along this line. When the cross-section increased in size he held it loosely, patted it gently, and even talked soothingly. As soon as it started to decrease, he screwed up his stick and bore down on it remorselessly. For an hour he wrestled. Then the professor entered with two 16 by 20 photographs taken out of frames.

"Wait!" shouted Heagey peremptorily. "Stand there and hold 'em." He twisted up his stick again, held it, and loosened it; and was rewarded by seeing the barrel-shaped mass appear; then two long, cylindrical bodies beside it, covered with metal-mesh.

"What's your idea?" the professor asked.

"Don't bother me!" Heagey panted irritatedly. "And don't move. I might need you any minute."

Finally the Thing decreased in size again; but this time Heagey seemed satisfied with it. He removed the canvas sack. There was an irregular sphere the size of a bucket. Over its surface were queer patches, glassy places, and iridescent, rainbowlike spots that changed color and looked deep.

"Quick now, the pictures!"

Heagey set up the pictures in front of the Thing, as if to show them to it. The professor stared at him as he would at a silly child. Heagey suddenly hit himself in the side of the head with his fist.

"What a prize fool! I keep on being a fool!" he shouted. He turned savagely to the professor.

"Get me the two best fellows out of the Fine Arts Department. Quick! Sculptors!"

If the professor thought Heagey was crazy, nevertheless some glimmer of hope of rescuing Sheila lent him willingness and speed of thinking. He scolded rapidly into the telephone for a few minutes, repeating the word "emergency" several times. Then he started down the driveway, taking a policeman with him.

Heagey was feverishly busy. He seemed to be bringing every object in the room that could be conveniently carried, to set before the unearthly specimen he had there. He seemed to be showing it things. He acted like some ignorant, superstitious savage, bringing things to his god. Books, chairs, hats and coats, mathematical medals, hammers and wrenches, one thing after another; he held them up in front of it for a while and tossed them aside on the growing heap. When the two sculptors arrived, he barked his directions at them, and continued what seemed his silly efforts to entertain the object in front of him by showing it everything he could find. At least it remained quiet and unchanged.

The sculptors, infected with his determination, worked rapidly. First there was a model of a heavy, bulging man, with his foot caught in a hole like a coal chute, and held fast by a square lid. Then from the pictures a model of Sheila; considering the speed with which it was made, it was a wonderful thing, with her pointed chin and curly hair all true to life. Then a rough model of Heagey.

Heagey set the models down in front of the iridescent, patchy Thing and played puppets with the models; went through a regular dramatic performance with them. The models of Sheila and himself stood near the man caught in the trap door. The imprisoned man struggled and knocked Sheila over and she rolled away; she fell down off the surface of the block to a lower level. The imprisoned man continued to struggle, and the model of Heagey searched around, but could not get past the edge of the block.

Then, very impressively motioning toward the Thing, as though he really believed it was looking, Heagey made the model of the imprisoned man lean over and pick up Sheila, and hand her over to the model of himself. The model of himself held on to Sheila, and raised the trap door that imprisoned the bulging man, who hopped out of the hole and hastened away. That was the little show that Heagey put on with the yard-high clay models.

The patchy sphere changed suddenly. First it shrank and then it swelled; then there were three or four Things moving back and forth. And suddenly, there stood Sheila!

Pale and distracted and wan she looked; and she swayed as she looked blankly around. Then her eyes widened and she gave a little scream; but a look of peace and content spread over her features. By the time Heagey was at her side, she fell limply into his arms.

"One moment, dear," he said gently as he laid her down carefully in the armchair. The professor was down on the floor beside her in a moment, watching her fluttering eyelids.

"Dad?" she breathed. "I'm all right."

Heagey stepped quickly to his switches and threw the big one in and out again. Again came the crash cut short, and the sensation of movement. And the Thing was gone. There was nothing left of it at all.

"Did you let the Thing go?" the professor reproved querulously.

"I had to," Heagey snapped. "It was a promise—for finding Sheila."

The professor was sitting on the floor, writing a check.

"Do you think you deserve this?" he said testily. He was merely trying to hide his emotion. "You won't get it until you prove it. Explain how you did this!"

Heagey dropped into a chair, looking exhausted to the limit.

"I reasoned from the things I saw Out There that this creature must be intelligent. There were buildings, machines, and leather and metal-webbing. So I made models and tried to deduce its shape. Somewhere on it there must be a head and eyes. You saw how I coaxed it 'through' this 'space' of ours until the head was cut by our 'space' and the eyes could see us. Then I told it what I wanted it to do with models—just as I would explain things to you by means of drawings on a sheet of paper.

"Now do you believe there are four dimensions?" Heagey demanded by way of vengeance.

"Hm. Do you?" the professor countered.

"Four? I'm convinced there are a dozen or a thousand dimensions!"

WILLIAM HAZLETT UPSON

# A. Botts and the Moebius Strip

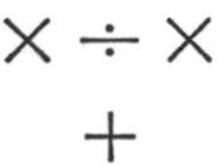

HOLLANDIA, NEW GUINEA,
Saturday, July 21, 1945.

MAJOR ALEXANDER BOTTS, AUS,
MUNGOMORI ISLAND.
DEAR BOTTS:

My friend Gen. E. E. Smith, of the Australian Army, with whom I have been conferring here on various tractor problems, tells me you have gone with the Australian Air Force to Mungomori Island, to conduct experiments in dropping tractors by parachute.

I am glad to inform you that Lieutenants Dixon and Humbolt, of the American Army, formerly safety engineers with the Earthworm Tractor Company, are due to arrive at the harbor on Mungomori next Tuesday, July twenty-fourth. They are making a checkup of Earthworm equipment all over the Australian area and suggesting measures for cutting down accidents and operational casualties. For this work they have been assigned a small escort vessel, and for demonstration purposes they have a sixty-H.P. Earthworm tractor and bulldozer, equipped with all the latest safety devices.

As former sales manager of our company, you should be interested in their project. I trust you may have a chance to meet them at Mungomori. And any courtesies you may show them will be appreciated by,

Yours very sincerely,
GILBERT HENDERSON,
*President, Earthworm Tractor Company.*

FIELD HOSPITAL NO. 334,
MUNGOMORI ISLAND,
Monday, July 23, 1945.

DEAR HENDERSON:

Your letter stating that a couple of guys are arriving tomorrow with a tractor brings a veritable burst of refulgent hope into a situation that has been fraught with nothing but black despair ever since day before yesterday.

On the morning of that day—Saturday—my Australian colleagues and I were over at our base on New Guinea preparing for the first test of our new and unprecedented mammoth parachute designed to drop a full-sized five-ton Earthworm tractor, complete with bulldozer, from a heavy bomber. Word reached us that a bulldozer was needed at once to make an air strip in a remote mountain valley twenty miles from the harbor on Mungomori, so that thirty Australian soldiers, severely wounded while cleaning the Japs from nearby caves, could be flown out. The men were in a poorly equipped field hospital. They had to be moved at once. Their condition was too critical to take them over the rough mountain trails. Air transport was the only answer.

We promptly decided to combine our parachute experiment with an errand of mercy. We made the long flight to Mungomori. We let go the tractor with its stupendous parachute over the valley. Two of my best tractor mechanics, Sergeants Venturi and Watkins, followed in standard chutes. I followed them, also in a standard chute.

The experiment was not a complete success. Venturi and Watkins landed safely. I escaped with nothing more serious than a sprained ankle and minor contusions sustained as I came down through the branches of a tree. But the big parachute, after breaking a few shrouds, sideslipped over into a swamp, where the tractor became so completely mired that it will take another tractor to get it out.

Since then I have been nursing my ankle in a bed in the field hospital here, and the two sergeants have done little more than run in circles. I have sent dozens of radio messages—with no results. There are no tractors available on this island. The New Guinea airbase has no more giant parachutes to drop us another tractor. Any plane big enough to carry a tractor is too big to land at the little airfield down at the harbor. I was told it will take two weeks to get a tractor here by boat. Every other lead has fizzled out. And the doctor in charge of the field hospital and the head nurse have kept pleading so piteously with me to do something, I have almost gone crazy at my own helplessness.

Then, this evening, came your letter—by air mail to the harbor and by jeep over the mountains to our valley. And never before, Henderson, have you written anything which has brought such tidings of pure joy to a group of suffering humanity. In my happiness I actually burst into song, and then announced the thrilling news that a beautiful Earthworm tractor would land at the harbor tomorrow. The doctors, the nurses, my two sergeants and all the patients who were well enough to be informed were at once lifted from the valley of despondency to the heights of new hope and felicity.

Tomorrow morning, with my ankle well bandaged and with a pair of handsome crutches for emergencies, I will drive down to the harbor in a jeep with Venturi and Watkins. We will bring back the tractor, pull the other machine out of the swamp and clear an adequate air strip in two or three days.

With heartfelt thanks for sending me the good news, I remain, yours, bathed in the sunshine of pure happiness.

ALEXANDER BOTTS.

MUNGOMORI HARBOR,
Tuesday evening, July 24, 1945.

DEAR HENDERSON:

This is to inform you that the sunshine of pure happiness mentioned in my letter of yesterday evening has suddenly been blacked out by the dark clouds of a new disaster so unexpected and so idiotic that you will scarcely believe it possible. I arrived at the harbor late this afternoon. I talked to Lieutenant Dixon, who is in charge of both Lieutenant Humbolt and the tractor. And, incredible though it may seem, and in spite of all my most frantic arguments, this egregious shavetail flatly refused to let me use his machine.

Unfortunately, I have no authority to take it away from the noisome skunk. But I naturally have no intention of taking his refusal lying down. I am therefore forced to resort to low cunning. And, as this may result in trouble later on, I want a little co-operation from you.

Such being the case, I will now give you a brief résumé of what is going on, and what I want you to do.

It was early in the evening when I first discovered this obnoxious Dixon, along with his apparently innocuous sidekick Humbolt, at one end of the camp here, in a pump house used to pump water from a spring to a tank on the hillside which supplies water to the camp. I was at once impressed by the man's revoltingly conceited personality. And when I learned what he was doing, I realized that his mind is as stunted as his self-importance is overgrown.

Paying no attention to my polite request for his tractor, he insisted on showing me a large poster which he had hung on the wall, and which expressed the theme song of his crusade. At the top were the words, ACCIDENT PREVENTION BRINGS VICTORY. In the middle were a lot of gruesome pictures showing an assorted group of morons getting their hands, hair and clothes caught in various gears, belts, presses and machine tools. Other careless imbeciles were getting in the way of trucks and tractors. Underneath was a poem:

*Every accident such as these*
*Impedes the war against the Japanese.*

"Very praiseworthy, I am sure," I said courteously. "But right now it happens that we have a group of badly wounded Australian soldiers—"

His only response was to drag me around to inspect the different units of the pump installation, while he expounded his safety ideas. He showed me an Earthworm Diesel motor in a small steel house, and a pump in another house about thirty feet distant, with a four-inch belt running between. Inside the motor house, the pulley and the belt are covered by a guard. This, according to Dixon, is good. Between the two buildings, the belt runs through a long wooden box. This also is good. But inside the pump house there is no protection for the pulley or for the belt—which comes in one hole in the wall and goes out another.

"This," said Lieutenant Dixon, "is very bad, especially when you consider how dim the light is in here. I have provided a guard for the belt and pulley. But the guard has to be removed to oil the pump. Somebody might leave it off. So do you know what I am going to do to provide additional safety?"

"If you are interested," I said, "in the safety and well-being of those wounded Australian sol—"

"Tomorrow morning," said Dixon, "I am going to paint the belt red."

At this, a corporal, who seemed to be in charge of the pump station, spoke up. "Won't that make the belt slip?"

"I will paint the outside, but not the inside. If I spill any on the inside, I promise you I will clean it off."

"Okay," said the corporal. "I won't be here myself. The tank at the other end of the line is full. We don't have to pump tomorrow. So I am getting the day off. But I'll leave you some paint remover in case of need."

"I will be careful," said Dixon, "and I hope you all realize how important this painting is. Industrial statistics have amply proved that painting dangerous moving parts in bright and contrasting color provides a very effective warning and substantially reduces the incidence of accidents."

"I have no doubt of it," I said. "And now I will tell you why I need your tractor."

I gave him the full story of the wounded Australians, substantially as I have set it forth to you. But the incredible dolt was so obsessed with the meticulous details of his safety campaign that all my arguments were completely lost on him.

"If I lent you the tractor," he said, "I should have to waste several more days here. As this pumping motor is the only piece of Earthworm equipment at this harbor, I have already spent too much time. I do not intend even to unload my demonstration tractor at this unimportant locality. I must pursue with all possible speed my vital mission of installing much-needed safety measures in connection with the many thousands of Earthworm units in other parts of this theater of operations."

At this, I opened up with everything I had—logic, pathos, drama, eloquence, sarcasm and even threats. He could see no point of view but his own. I tried to pull my rank. He knew I had no authority over him.

His final words were, "Lieutenant Humbolt and I will spend the night on the boat. We will paint this belt first thing tomorrow. And by the middle of the morning we will sail for our next port of call."

Taking Humbolt with him, he departed.

Having failed with this Dixon person, I promptly took all the other more obvious measures. I tried to appeal to the local Australian commander. He was away on a trip up the coast, and nobody else wanted to assume the authority of interfering with an American lieutenant traveling under independent orders. I tried to send messages to the higher command and to you. But the pitiful little local radio station had suffered a breakdown.

Temporarily baffled, I hobbled back on my crutches to the barracks where I had been assigned a bed. I sat down to meditate. In my usual keen manner, I analyzed the situation. The tractor was on the boat. Probably I could bluff the captain of this craft into letting me take it the next morning while the two lieutenants were at the other end of the camp painting the belt. But this painting job would hardly take long enough to give me time to get the machine far enough on its way to prevent the lieutenants from catching me.

Was there any way to prolong the painting job? I remembered that the belt ran through two holes in the pump-house wall. It would be impossible to remove it without unlacing it—which would be too much trouble. The two lieutenants would probably paint it in place. One guy would probably crank the engine very slowly, causing the belt to travel past the other guy, who would apply the paint at the only place where the belt was accessible—in the pump house. I tried to picture all the details in my mind. Dixon had said they were going to paint one side of this belt, but not the other. I considered this point for some time. One side, but not the other—

And then, in a sudden brilliant flash of pure inspiration—Eureka! —I had the answer. I limped out and got hold of Sergeants Venturi and Watkins. I took them to the deserted pump house. We unlaced the belt, gave one end a half twist, and laced it together again—thus neatly converting it into that baffling mathematical paradox known as the Moebius Strip.

The effect this is going to have on the unsuspecting Dixon tomorrow morning will be something to see. When he tries to paint the outside without painting the inside, he will, of course, be attempting the impossible—because the belt now has only one side. If he persists in his ill-omened plan, he and his partner will be delayed indefinitely, and there will be plenty of time to get away with the tractor.

Having prepared my erudite little trap, I dismissed Venturi and Watkins, and returned to the barracks, where I have been writing this letter. It will go out by plane early tomorrow morning, and should reach you before noon. As soon as you receive it, I want you to bear down on your friend, General Smith, and make him send an order to Dixon directing him to let me have the tractor as long as I want. Make it retroactive, because I shall already have taken the machine, and the only need for the order is to prevent this mug from preferring charges against me or rushing over the mountains and trying to take back the tractor before we are through with it.

If you work fast, the order should be here tomorrow night, which ought to be in plenty of time. As my ankle still bothers me, I will stay here at the harbor and check up on the expected droll antics of Dixon and Humbolt, while Venturi and Watkins take the tractor over the mountains and clear the air strip.

Certainly, this peculiar belt-painting project has proved a rare piece of good luck for me. And, if my luck holds, I should be able, before the end of the week, to report complete success in the one mission which right now is uppermost in my mind and heart—the rescue of those gallant Australian soldiers.

Yours,
ALEXANDER BOTTS.

HOLLANDIA,
Wednesday, July 25, 1945.

DEAR BOTTS:

Your letter is received, and I agree with you emphatically that the rescue of the Australian soldiers should take precedence over Lieutenant Dixon's plans. General Smith is of the same opinion, and orders have been sent Dixon to place the tractor at your disposal.

I am worried, however, at your wild and irresponsible action in stealing this machine—even though it is in a good cause. And I deplore your talk of your "rare good luck." Someday, if you keep relying on mere luck to get you out of trouble, you will be sadly disappointed.

Incidentally, your garbled remarks about this Moebius Strip—whatever that may be—are obviously unsound. After all, a belt running around two pulleys is a simple contrivance, and no amount of pseudo-mathematical jargon and hocus-pocus can alter the fact that such a belt must have two sides. If a man wants to paint one side and not the other, there is no reason why he cannot do so.

Perhaps you are a little delirious from your parachute injury. Or maybe you have been overexerting yourself. In any case, I would strongly recommend that you take the earliest opportunity to get a complete and well-earned rest.

Yours very sincerely,
GILBERT HENDERSON.

MUNGOMORI HARBOR,
Saturday, July 28, 1945.

DEAR HENDERSON:

Thank you for your letter, which has been delivered to me here at the little Mungomori Harbor hospital, where I was dragged in last Wednesday in an unconscious condition brought on by the unexpected results of my Moebius Strip project, and where I am following your recommendation to take a well-earned rest. I also thank you for your promptness in getting through that order to Dixon.

But I must take exception to your erroneous belief that I get out of trouble by what you call "mere luck." Actually, there is nothing "mere" about it. I always earn my good luck.

In the present case, it just happened that this Dixon, whom I had to keep busy while I stole the tractor, was planning to paint one side, and one only, of a belt which just happened to be covered through most of its length. This was a lucky break, but only because I knew how to take advantage of it. An ordinary person like you would have been helpless; your letter, in fact, shows you are still all at sea, even after I wrote you my plans.

What I did was make use of the Moebius Strip, which I read about some years ago in a very fine book called *Mathematics and the Imagination*, by a couple of smart cookies named Kasner and Newman. As my previous description of this Moebius Strip failed to penetrate your somewhat obtuse mind, I would suggest that you try again with a model.

Get yourself a strip of paper about a yard long and an inch wide. Lay this flat on your desk, lift the two ends and bring them together. You now have an endless paper belt with two sides—an inside and an outside—similar to the ordinary belt I found at the pump house.

Next, separate the two ends, turn one of them over and bring the two ends together again. This time you might as well lap these ends and paste them. You have now done exactly what I did to that belt. You have created a Moebius Strip, whose outstanding feature is that, although any section of it obviously has two sides, it, nevertheless, possesses, in its entirety, only one side. Incidentally, it has only one edge. And you can get other interesting effects if you try to cut it into narrower strips lengthwise. But I do not wish to strain your intellect unduly.

The important thing, from my point of view, was that I had fixed that belt so that the outside was continuous with the inside, and so that Dixon, when he tried to paint the outside, but not the inside, would actually be attempting to paint—and at the same time not to paint—the same side. This, I hoped, might cause a certain delay in his operations.

As it turned out, my plans worked exactly as I had anticipated—for a while. Early on Wednesday morning, Venturi, Watkins and I repaired to the neighborhood of the pier and waited till Dixon and Humbolt had left the boat and were well on their way to the pump station a half mile away at the other end of the camp. Then we approached the officer in charge of the boat. With nonchalant assurance I explained that my men had orders to take the tractor. Having no reason to suspect dirty work from an American major, the officer acquiesced, and obligingly assigned some men to help unload the machine and place it on the pier. As this looked as if it would take at least an hour, I left Venturi and Watkins on their own, and ambled along on my crutches to the pump house.

As expected, I found Dixon standing next the pump pulley, applying a handsome coat of red paint to the upper surface of the belt. As soon as he finished as much as he could get at, he would let out a yell, and Humbolt, over in the motor house, would crank the belt along, so as to bring another section in reach.

"Good morning," I said.

"Good morning," he replied gruffly. "It won't do you any good to keep asking for that tractor—if that's what you came for."

"No," I replied. "I just stopped in because I am so deeply interested in your safety campaign. . . . Look out! You're going to spill paint on the inside of the belt."

"I certainly am not."

"If you do, I hope you remember you promised to clean it off."

"I remember."

I limped outside and took a look in the direction of the pier. Apparently the tractor had not yet got going. After a while I went back into the pump house. Dixon was going faster than I liked.

"Here is something funny," he said. "I thought I painted past the belt lacing awhile ago, and here it comes again."

This took a bit of quick thinking on my part. I realized that the belt lacing actually had been past once before, at which time the guy had painted the other side of it. If he started investigating, he might find out too much too soon.

"Probably," I said carelessly, "the belt is made up out of two or three pieces, so there are two or three lacings."

"Maybe so," he said, and went on wielding his brush.

More time went by. I went out and had another look toward the pier. There were no signs of the tractor. I thought of the desperate plight of the Australian soldiers. I began to be worried. I returned to the pump house.

The painting went on—interspersed with yells from Dixon, followed by short pauses while Humbolt, at the other end, cranked the belt along. Finally there emerged a section already painted red. Dixon covered the remaining surface, and stepped back to admire his work. The job was finished.

I moved forward on my crutches. "Too bad!' I exclaimed. "You got some paint on the inside."

"I did not! At least, I don't think I did."

"Look."

Dixon bent over in the dim light. He ran his finger along the inner belt surface. It came away dripping red. "I can't understand it," he said. "I took the greatest pains not to spill a single drop. But the inside is completely covered as far as I can see."

"You don't suppose," I asked incredulously, "that the inside is painted solid all the way around?"

"Hey, Humbolt!" yelled Dixon. "Crank this thing along! Keep going till I tell you to stop!"

Slowly the belt began moving. We watched. It went around several times. Sure enough, the entire inner surface was painted a rich and beautiful red.

"It's impossible!" said Dixon.

"You certainly must be absent-minded," I said cheerfully. "Imagine just painting along—here and there, on one side and the other—without paying any attention to what you are doing."

"You know I didn't do any such thing," he replied angrily. "Somebody has been slipping something over on me."

"Possibly gremlins."

"Don't be silly." He gave me a keen look. "Say," he said, "you didn't have anything to do with this, did you?"

"How could I? I haven't been near that belt since you started."

"Maybe this was done before I started."

"It's your own paint—same shade and still wet on both sides. If you're trying to blame it on somebody else just so you can crawl out of your promise to clean it off—"

"I'm not trying to crawl out of anything! But I tell you I didn't paint the inside of this belt!"

"You must have—unless maybe this Humbolt is a practical joker."

"What do you mean?"

"Well, he could have brought along a can of this paint himself, couldn't he? While you were painting the outside from this end, he could have been painting the inside from the other end—just for a little joke, perhaps."

"Why, the big bum!" said Dixon. He went rushing out the door and over to the other building. I followed as fast as I could, and arrived just in time for his opening blast.

"Come clean!" he snarled. "Why did you smear up this belt?"

Humbolt looked mildly astonished. Dixon advanced threateningly. "Did you or did you not paint the inside of this belt?"

"I did not," said Humbolt. "How could I, when it's all covered up by that guard? And I have no paint anyway."

Dixon inspected the guard. It was firmly fastened in place by at least a dozen bolts. None of the nuts looked as if they had been recently tampered with.

"If you would tell me what is the matter," said Humbolt, "maybe I could be of some help."

"Come with me," said Dixon grimly.

He led his assistant out the door and back to the pump house. As I straggled along in the rear, I reflected that never before had I heard a conversation so inane, and yet so happily contrived to use up a lot of time that needed so badly to be used up in just such a way as this.

Back in the pump house, the two lieutenants examined the belt in bewilderment.

"The paint remover," I said, "is right there in the corner."

"All right, all right," said Dixon wearily. "You may go back to the motor house, Humbolt, and stand by to crank this thing when I give the word. And remember, no funny business."

Humbolt departed, and Dixon started the long job of scrubbing the paint from the inside of the belt. It soon appeared that the paint remover, which is designed to soften up old dried paint, was not much help here. It only made the wet paint wetter. Dixon wandered around the pump house and gathered in all the old rags and cotton waste he could find. He got a pail of Diesel fuel, and when this did not work too well, he brought another pail filled with gasoline from the small starting motor on the side of the Diesel. He finally sat himself on the floor beside the pulley, and settled down to rubbing, scrubbing and wiping his way along the inner surface of the belt, stopping only long enough for Humbolt to crank the thing along.

After the poor man had slaved for a good while, I wandered outside and almost ran into the mail orderly, who was heading for the nearby barracks where I had been assigned a bed. He handed me a letter. It was from the head nurse at the advance field hospital, begging me not to delay in getting back with the tractor. Several of the patients were worse. It was more than ever necessary to fly them out at once. They were counting on me to prepare the landing strip, so the planes could come in. I was their only hope.

Anxiously I gazed across the camp toward the pier. I strained my ears. From far away came the beautiful sound of an Earthworm-tractor motor. Then I saw the machine itself, just leaving the neighborhood of the pier and starting along the trail that led over the mountain. My heart beat fast with hope. But the tractor had taken a long time to get started. It was moving with maddening slowness. And it would probably take another half hour to reach the jungle, where it would be safely out of sight. If Dixon saw it before it disappeared, he might still have time to get together his men on the boat, organize a pursuit and bring the machine back.

I re-entered the pump house, where I was dismayed to find that Dixon was getting along entirely too fast. That portion of the belt's surface which he had first scrubbed clean on the inside had now been cranked all the way to the other end, then back again, and around the pump pulley—this time on the outside. The unsuspecting Dixon, seated on the floor with his head lower than the upper part of the belt, had not, as yet, observed this interesting phenomenon—which seemed to me just as well. The longer I could keep him working along quietly, the better it would suit me.

Unfortunately, however, the man was restless. Hearing my step behind him, he turned his head and began complaining that the job was taking too long. "Somewhere," he said, "I have heard that the quickest way to remove paint is to burn it off with a blowtorch."

"That is sometimes done," I admitted.

"And isn't that a gasoline blowtorch over there on that workbench?"

"Yes, but a blowtorch is used only to remove paint from something solid, like structural-steel work, or sometimes, with proper precautions, from wood. You can't use a blowtorch on that belt. You would burn the material."

"Not if I'm careful," he said. "I'm going to try it anyway."

He rose—with never a glance at the top of the belt—gathered up his rags, his cotton waste and his pails of fuel oil and gasoline, and carried them over and set them down beside his paint cans in the doorway of the building. He then stepped to the workbench, pumped up a bit of pressure in the blowtorch, opened the valve and lighted it.

"You can't do that," I said, "that belt is soaked with gasoline."

"You may outrank me, Major Botts," he said coldly, "but I am not aware that you have any authority over me in this matter."

He picked up the torch and walked toward the pump. He stopped. He pointed.

"Look!" he said weakly.

"What's the matter now?" I asked cheerfully.

He spoke haltingly. "The paint . . . has disappeared . . . from the outside of the belt."

"So it has, so it has!" I exclaimed, in tones designed to express surprise. "How did that happen?"

"I don't know—yet," said Dixon. His voice was getting ugly. "But I'm going to find out. And when I do, somebody is going to be sorry." He began to shout, "Humbolt! Humbolt! Come here, right this minute!"

A half minute later, Lieutenant Humbolt joined us. "Did you want something?" he asked.

"Yes, I want to know what sort of monkey business you think you are pulling off now."

"Monkey business?"

"Yes. Did you or did you not scrub the paint off the outside of this belt?"

"I thought I was just supposed to crank the engine over. Was I supposed to be scrubbing paint too?"

"You were not. But somebody has been taking the paint off the outside of this belt."

Humbolt took a look. "That's right," he said. "But what's the idea? I thought you were just going to take it off the inside. Why did you take it off the outside too?"

By this time Dixon was ready to blow up.

"I did not take it off the outside. But somebody did. And I'm going to find out who it was."

"Well, I didn't."

"You deny it?"

"Of course I do. And, furthermore, I don't see why you keep accusing me of messing up your old paint job. First you claim I painted the inside. Now you claim I cleaned off the outside. Probably you did it yourself."

"That will do, Lieutenant Humbolt. If you don't want to co-operate in this matter, I am perfectly capable of taking care of it myself. Before we proceed any further, I am going down to the boat, and I will bring back an adequate guard to watch both ends of this job."

"Oh, I wouldn't do that," I said hastily.

"Why not?" he demanded.

For a moment I hesitated. The real reason why I did not want him going down to the boat was that I did not want him to discover the departure of the tractor until it had had time to get a safe distance away. Naturally, I could not tell him this. But I had to say something, if only to keep up the conversation and use up a little more time. I decided to speak softly.

"If you ask me," I said, "the explanation is very simple. It is nothing to get excited about—just a little innocent absent-mindedness on your part, Lieutenant Dixon. And what a droll performance it has been, to be sure! First, while painting the outside, you absent-mindedly cover the inside. Then, while cleaning the inside, you absent-mindedly scrub off the outside. The next you know, you'll be giving yourself a shampoo with the paint remover or possibly painting up the inside of your pants."

Unfortunately, this type of soft answer did not seem to turn away any wrath.

Lieutenant Dixon became what I can only describe as livid with rage.

"That's a lie, and you know it!" he yelled. He took a couple of steps toward me and swung back the blowtorch, preparing, apparently, to bean me with it. This was a mistake on his part. In his excitement, the torch slipped out of his fingers, sailed back across the room and landed in the pail of gasoline. There was a flash and a roar. The doorway—which was our only means of escape—was at once a mass of flames.

The three of us retreated to the other end of the building. Right away the whole place filled with thick oily smoke from the burning gasoline, fuel oil, rags and paint. We dropped to the floor, where the air was clearer. We crawled behind a pile of boxes. The smoke swirled down around us. We couldn't get away from it. I felt myself choking . . . choking . . . choking——

When I came to, I was in the hospital. That was three days ago, and by this time I have learned what happened. Some Australians arrived and put out the fire in a hurry with some sort of fancy foam-extinguisher equipment. The building, the pump and even the belt were saved. Dixon, Humbolt and I were merely overcome by smoke. We are all right now, although I am staying in the hospital till my sprained ankle gets better.

Venturi and Watkins crossed the mountain, jerked the other tractor out of the swamp and cleared a primitive air strip—all in two days. Today, the Australian wounded were flown down here in small planes and sent on toward Australia in big seaplanes. Venturi and Watkins hauled the hospital equipment out over the trail and returned Lieutenant Dixon's tractor. And the head nurse from the field hospital has just been in and wept tears of gratitude on my shoulder.

You may also be interested to know that the stubborn Lieutenant Dixon was all for going back and finishing the impossible task of making that one-sided belt red on one side but not on the other. However, the Australian colonel in command of the port, who got back just in time to see the fire, has ordered both Dixon and Humbolt on their way. "These safety engineers," he says, "may be all right back home in peacetime, but over here where there is a war going on they are entirely too dangerous."

Yours,
ALEXANDER BOTTS.

NIGEL BALCHIN

# God and the Machine

× ÷ ×
+

"APART FROM causing some fifty thousand deaths," said my Uncle Charles, "the first atomic bomb has much to answer for. It not only blew up Hiroshima. It also blew up the egos of some of our scientists to a remarkable size." He tapped his morning paper. "I read here a speech by Professor Plumreach, in which he explains that the world will come to no good unless scientists demand their proper place in executive control of practically everything. Reading between the lines, I should guess that Plumreach himself would be willing to consider taking over the Premiership, combining it, perhaps, with the Presidency of the United States."

"I don't think I know Plumreach," I said. "Is he an atomic bombardier?"

"Not exactly," said my Uncle Charles. "Through some oversight, he does not appear to have been employed on the project itself. But he has made a comfortable living for the last few years writing about it as though it was his personal property." He leaned back in his chair and closed his eyes. "I don't know what it is about the physical sciences," he said rather plaintively. "But they sometimes produce a certain type of stupid pomposity quite unequaled in any other job except that of a small-town mayor. The botanist, the zoologist, the physiologist—even, God help us, the doctor—usually retain some shreds of humility. But the physicist, who often has barely enough sense to come inside when it rains, nearly always feels that anybody who doesn't understand his particular set of parlor tricks is a childish barbarian."

"Still," I said, "one has to admit that the parlor tricks can be pretty impressive at times. The electronic computer, for example."

My Uncle Charles sighed. "You should know by now," he said, "that it is quite useless to talk to me in technical phrases. What *is* an electronic computer?"

"I can't say I really understand it very well myself. But it's the thing that the press have been calling the Mechanical Brain."

My Uncle Charles nodded. "Ah," he said. "So they have invented that again? I assume that it can do sums, shake hands, say 'Hallo' and sing a restricted range of songs? I believe that is the normal repertoire. And not a thing in it that I cannot do myself, up to a point at least."

"I gather that the sums are its strong point. It can carry out very complicated calculations at a terrific speed. Apparently it can do in a few seconds jobs that would take a team containing every mathematician in the country fifty years."

"The last one I saw," said my Uncle Charles coldly, "could shake hands. I personally never shook hands with it, but I saw others do so. Surely there have been mere calculating machines for many years?"

"Yes, but the point about this one is that it has a sort of 'memory.' It can store facts for use later—which means that instead of just doing one calculation and finishing, it can do a calculation or a whole series, have a look at the results, and then decide what to do next."

"Fascinating," said my Uncle Charles wearily. "The last one I saw sang—though not well."

"It can do two-move chess problems."

"Once again, so can I, but nobody makes a fuss about . . ." My Uncle Charles stopped suddenly and opened his eyes. "One moment," he said in a different voice. "Does it also play draughts?"

"Yes."

"You should have said so before. If it plays draughts it is clearly a development of the work of Dr. Scouler, in which I was to some small extent involved."

"I don't think I know about Dr. Scouler."

My Uncle Charles shook his head. "You wouldn't," he said gently. "But there is no reason why you should not do so."

"Having heard my opinion of physicists, you may find it surprising that Dr. Scouler should have been a friend of mine, for he was not only a physicist, but a physicist of the most virulent type. Indeed, that was what first brought us together, many years ago. Scouler was a member of my club, and from time to time, in the bar, the smoking room, or wherever two or three were gathered together, he would produce samples of this arrogant scientific nonsense to which I have been objecting. My club is a plain-spoken place, and on these occasions I never hesitated to tell him that he was being a pompous ass.

"But Scouler was of that curious type which never distinguished between an old friend and an old enemy. One had only to be rude to him half a dozen times, and one became, if not a friend, an acquaintance to be sought out and talked to. I don't think he dealt in friends. He was not popular, though he had the reputation, in his own field, of being a very brilliant man.

"This was as long ago as in the twenties, and I cannot now recall how I came to visit Scouler's home. But I remember that it made a very unpleasant impression on me. He was a comparatively young man at the time and only just beginning to make his mark, but he was married and had two children—a boy and a girl. I should think the household was pretty short of money—it certainly gave that impression. But what I found unpleasant was Scouler's attitude to his family, which was a magnified edition of his silliest efforts at the club—patronage and arrogance to an extreme degree. He referred to his wife, and treated her, as though she was a moron, and more or less invited one to do the same. She was clearly terrified of him. With the children he had a sort of heavy teasing manner, which took every simple thing they did or said and made it the subject of complicated pseudo-scientific argument, in which the main object seemed to be to embarrass them and make them look fools. I remember that on one occasion the boy, who was about ten, spilled some water. Instead of taking no notice, or calling him a clumsy little ass, Scouler at once launched into a dissertation about the physical properties of fluids, talking ostensibly to me but at the child, and punctuated with 'As Roy is well aware' and 'As my son's studies will doubtless have taught him.' In the end the boy burst into tears, which seemed to give Scouler pleasure.

"I found all this very embarrassing and unpleasant, and never went to the house again. Indeed, after that, I rather avoided Scouler; and when, a few months later, he left London and went to take up a post in one of the provincial universities, I think the club was glad to be rid of him.

"I did not see Scouler again for years, though I heard of him occasionally. The brilliant career in his particular field was duly taking place, and I believe he was regarded as one of the outstanding experimenters in the country. Then one day somewhere round about 1937 he turned up again at the club. He had not changed very much, either in appearance or manner, except that his belief that mankind consisted of an aristocracy of physicists and a rabble of others seemed, if possible, more marked than ever.

"He was working in London again and was staying the night at the club where, at that time, I was a resident. It was therefore inevitable that late at night, when the place had emptied, I should find myself in conversation with him, and I happened to ask after his family. As soon as I mentioned them Scouler's face took on a curious rigid look that might almost have been one of hatred. He said curtly, 'That's a thing I prefer not to discuss, if you don't mind.'

"There was nothing in the world I wanted to talk about less than Scouler's family. I said, 'I see. I'm sorry . . .' and was about to launch hurriedly into some other subject. But he was too quick for me. He said, 'You saw them once—all three of them. You probably had an inkling of what was coming. But I, of course, blind with affection . . .'

"He talked about it for two hours, and would probably have gone on till dawn if I had let him. In many ways it was a very ordinary tale of woe. About five years after I had seen him last, his wife had left him. Presumably even the meekest and most frightened worms have some turn left in them. The boy, much against his will, had been sent to Cambridge to read physics, had been sent down for idleness and drunkenness and was now a shop assistant. The girl, who had been intended to read chemistry at London, had demanded at the age of eighteen to marry somebody whom Scouler thought unsuitable, and on permission's being refused had run away with the man. Scouler did not know the exact whereabouts of either of them.

"The only remarkable thing about all this was Scouler's own attitude. It clearly never entered his head that he was in any way to blame. He was simply a man who had been done deliberate and calculated wrong in return for good. Time and again he used the expression: 'I have been cheated,' and I gradually realized that he was using the expression in the same sense as a man would use it if he had been deliberately given false change. In his view, the fact that he had married his wife and supported her, and begotten his children and brought them up, not only gave him a claim on them, but a *mathematical* certainty of their obedience and affection. The fact that they did not agree with this equation was not only an offense against him, but an offense against some basic truth, as though they had argued that two and two made five.

"I naturally listened without comment. There was no useful comment to make. And then gradually Scouler left the subject of his family and started to talk about his work. The breakup of his home, he explained, had had one good result; it had left him freer, in every respect, to concentrate on research and experiment. He implied that he had practically withdrawn from the world into the laboratory, and found the exchange a good one. What I think he was trying to say, poor devil, was that he found electrons easier to understand and to handle than human beings. But his arrogance and sense of injury would not allow him to put it in such simple and understandable terms. The way he liked to think of it was that physical phenomena had about them the original and unsullied truth and honesty that he had found lacking in mankind. 'In my book,' he said, 'I may make mistakes. I may deceive myself. But at least nothing sets out to deceive me. The truth is there, if I can find it, and it is the same truth always for everybody. It does not alter and shift its ground for some petty personal reason. Deliberate cheating and lying, thank God, are phenomena confined to the human species.'

"This struck me as irritatingly childish even for Scouler. 'Oh come,' I said, 'you're rather suggesting that the difference between a man and any other sort of matter is that one cheats and the other doesn't. A man may tell lies and a brick may not. But if it comes to that, bricks don't write lyric verse and play violins. Inanimate matter may be superbly truthful, but it's uncommonly dull company in a pub. One has to pay something for the advantages of human intelligence.'

" 'Possibly,' said Scouler, without enthusiasm. 'But I think one often has to pay too much. Particularly as a lot of people's respect for "human intelligence" is the product of mere ignorance. You mention violin playing for example. It would be perfectly simple, given the time and the money, to devise a mechanical violin player that . . .'

" 'Oh I dare say. And a mechanical sonnet writer. But the thing couldn't *think*, and it couldn't feel emotion, could it?'

" 'Much depends,' said Scouler, in his most superior way, 'on the exact sense in which you use the words. Probably, like most people, you attach no exact meaning to them. But whereas there are some processes of the human brain that cannot at present be reproduced scientifically, they become rapidly fewer as science advances. The irresistible conclusion is that "human intelligence" is only a convenient phrase for certain mechanical processes that will eventually prove capable of analysis and reproduction.'

" 'You look forward to a machine that will write *Hamlet?*'

" 'I see nothing intrinsically absurd in the idea.' Scouler paused. 'Do you play draughts?'

" 'I have done so.'

" 'Would you say it involved the use of the human intelligence?'

" 'I suppose so. To a mild extent.'

" 'To quite an extent surely? It involves knowledge of a set of rules, ability to make choices and decisions, and so on.'

" 'Yes.'

"Scouler smiled. 'Nevertheless if you come to my laboratory one evening you can play a game of draughts with a machine on which I am working, and if you care to have five pounds on the result I shall be happy to take you.' He held up a hand. 'Don't tell me that there is a big difference between playing draughts and writing *Hamlet*. I know it. But you must give us time.' He smiled. 'After all, your beloved human intelligence has several thousand years of experiment and development behind it, hasn't it?'

"A week or so later I visited Scouler's laboratory, which was in a rather unexpected place out on the Great West Road. It was a brand-new laboratory, and I gathered that it had been specially built for his department.

"I had visualized his draughts-playing machine as being the conventional idea of a robot—a sort of cross between a suit of armor and a cash register, that would put out a rather jerky steel arm and move the men on the board. In fact what I found was a complete roomful of machinery, vaguely reminiscent of a small power station.

"I said, 'This is a very massive affair, Scouler. I had been thinking that a machine that could play draughts might be a pleasant companion in the house for the long winter evenings. But I see that few of us could accommodate it. How much do they cost?'

"Scouler said, 'This one has cost about £50,000 so far. But of course it's only in its infancy. It will cost at least another hundred thousand before I'm through.'

"Privately, I felt that even fifty thousand was a lot of money to get oneself a game of draughts, but I did not say so. Scouler then proceeded to tell me about the machine."

My Uncle Charles paused and wrinkled his brow. "You will not expect me to remember what he said. It was all far beyond me—particularly since Scouler took a delight in delivering these technical lectures in a way that nobody but an expert could possibly understand, while pretending to think that one was following him with ease. There was plenty of 'As of course you know' and 'As you will recall,' and I was vividly reminded of the wretched small boy when he spilled the water. But I do remember that he said that the machine had to be given what he called 'a program' which was a set of basic instructions, and that it then played the game by considering every possible move and deciding on the best one by a series of very rapid mathematical calculations. He pointed out, which I found interesting, that it would be equally possible to instruct the machine to lose. But in this instance its 'program' was to play perfectly, which meant that even if I played perfectly too, I could never do more than draw, and if I ever made a slip, must inevitably be beaten.

"He had started off on this disquisition in his usual cold, rather supercilious, lecturer way. But as he spoke of the speed, perfection and completeness of the machine's calculations his manner changed. His eyes began to shine with enthusiasm, and he spoke as one sometimes speaks of Piero, or a fine claret, or even of God—as though, in fact, he was rejoicing, and asking one to rejoice with him, in some glimpse of ultimate truth and beauty. I must say I found him more attractive like this. I am not an enthusiast, and certainly not about mathematical calculation; but I have a weakness for enthusiasm in others.

"And then quite suddenly his mood changed, and for no obvious reason he began to talk about his family and how it had 'cheated' him.

It had been bad enough when he had done so before at the club, but he spoke of them now with a bitterness and hatred that was very unpleasant. I tried to point out that I had come there to play draughts with his machine, and not to listen to abuse of his family, but he took no notice. And gradually I realized that the two issues—the perfection and beauty of the machine and the imperfections and ugliness of his wife and children—were closely connected in his mind. He was contrasting them, for my benefit and his own. This roomful of machinery was the perfect child of his perfect second marriage—his marriage to his work.

"This must have gone on for quite half an hour, and I was beginning to think that I should just have to make some excuse and go, when he stopped just as suddenly as he had started and suggested that I should play a game with the machine.

"The 'board' consisted of an illuminated panel in the front of the machine, on which the men appeared as red and white lights. In front of me was a reproduction of a board with press buttons in each square, and I made my moves by pressing the button in the square to which I wanted to move, which automatically altered the position of the lights on the panel. The machine's next move, which was made practically instantaneously, then appeared as another alteration of the lights —and so on. When a man was taken, the light representing it went out, and if it became a King, it became larger. It was ingenious, but a little awkward when one was unused to it. Moreover, I had not played draughts for some time. In the first two or or three games we played I made a slip and the machine won with ease. The games gave me a curious sensation of helplessness, increased, I fancy, by the speed at which the machine's moves followed one's own. If I paused for a moment to consider a move, I could almost hear the thing tapping impatiently with its foot.

"As I think I have told you before, I am not really a very good audience for scientific marvels. I seem to be markedly deficient in the sense of wonder. It might truly be said of me that a miracle by the river's brim a simple miracle was to him, and it was nothing more.

"In the present instance, after playing a few games with Scouler's machine I was perfectly prepared to admit that it could play draughts, and that, to me, was the end of the matter. I had no desire to go on playing with it, since I found its personality unattractive. Apart from seeming impatient, it lacked the lighter touch, which to me is an important part of the game of draughts. But Scouler was obviously getting intense pleasure out of its performance, and held forth lyrically about its speed, skill and precision. He never mentioned my skill or precision, so that I felt rather as a footballer must do when playing away from home in a cup tie. However, he insisted that we should go on and on, and it was mainly through boredom that I eventually decided to try an experiment.

"I had noticed that while the machine played quite perfectly, and pounced at once on any mistake of mine, its perfection lay purely in *response*. It never laid traps for me or did anything unexpected, and after a while I began to suspect that it did not really understand the game very well, and might be disconcerted by unorthodoxy.

"I therefore began to make moves which were not so much wrong as irrelevant, and though sometimes they turned out to be fatal, and were immediately pounced on, at others there was a distinctly long interval before the machine's next move, and that sometimes seemed irrelevant too.

"Scouler noticed this, and launched into a long technical explanation, the gist of which was that this was not the fault of the machine but because the 'program' it had been given was one for simple demonstration against a reasonable opponent and made little provision for what he called 'fooling about.' He was rather sniffy about it, with a marked implication that I was not behaving in a sporting way. But by now I was interested, and after a while I succeeded in playing a game so peculiar that the machine seemed to become completely lost and threw away its men right and left, so that I obviously had won a game. Scouler had stopped explaining now and was sitting quite silent, staring at the illuminated board. There was real pain and loss in his face, and for a moment I heartily wished I had not made the experiment. But there was nothing to be done now but to go on, so I made the move which should win me the game."

My Uncle Charles paused and stared reflectively at his fingers. "There was an interval of some seconds," he said carefully, "which was much longer than any interval before one of the machine's moves. And then it cheated. Absolutely clearly, without apparent shame, it simply moved a man backwards—the sort of pathetically transparent cheating that a small boy might have tried in the same circumstances. I think I laughed and said, 'Oh come!' or 'Hey . . .!' or something like that. And then I saw Scouler's face. He had gone a queer whitish-gray color and he was staring at the machine with a sort of startled horror, like a man who has just seen murder done. He stood there for some seconds in silence, and then turned and went out without a word. He never looked at me.

"I waited for a few moments to see if he would come back, and then went to look for him. He had gone out of the building to his car, and was already starting the engine. It was getting late by now, and, as I have told you, we were somewhere out on the Great West Road. Having no desire to be abandoned with no means of getting back to London, I called, 'Hey—wait a bit' and got in beside him. He neither looked at me nor spoke, but simply started to drive.

"We must have driven for ten minutes in complete silence. Then I said, 'That was very interesting.'

"Scouler hesitated and then said, 'It cheated.'

" 'Yes. But not very well. You should train it to kick the board over in those circumstances.'

"After a while Scouler said wearily, 'It's quite simple of course, really.'

" 'Of course it is. It didn't want to lose.'

" 'In putting in the program,' said Scouler, as though he hadn't heard me, 'one prohibits it, of course, from breaking the rules of the game, and also from losing . . .'

" 'Most of us have programs like that.'

" '. . . but—*but*, there is, of course, no such thing as an *absolute* prohibition, because the thing depends on numbers, and there is no infinite number. So all you can do is to prohibit losing and cheating up to the maximum number of digits the machine can handle. So faced with two incompatible prohibitions it works on till its capacity is exhausted and then . . .'

" 'Then it cheats?'

" 'No,' said Scouler dully. 'Not necessarily.'

" 'It might equally well lose? That was equally prohibited.'

" 'No. When it gets to the last digit it records whether it is odd or even. If it is odd it breaks the prohibition against losing. If it is even it breaks the prohibition against cheating. That's all.' He was silent for a moment and then said almost fiercely, 'What else can you do? You've got to give it a program, and there simply isn't any way that you can put in an *absolute* prohibition.'

"I said, 'The moralists have encountered a similar difficulty. Frankly,

Scouler, I thought that touch of original sin was the thing's most endearing gesture.'

"There was a long pause and then suddenly he gave a sort of long crow of laughter. 'Yes,' he said, 'I suppose it was rather funny—if you look at it like that.' "

My Uncle Charles crushed out the stub of his cigar and rose. "It is now not only five-thirty but five thirty-one," he said, "and we should not waste valuable time. . . ."

"But what happened to Scouler and the machine?"

"I really don't know. I imagine that the machine was a crude early version of this new thing you mention. Scouler may have developed it further, or he may not. He certainly did very valuable work in the war on radar and so forth. He was decorated for it and died a couple of years ago. I never saw much of him after that night. In fact I can only remember one occasion. He was dining in a restaurant with his son and daughter and introduced them to me. They seemed a nice pair, and I must say Scouler himself seemed to have mellowed. It is now five thirty-two."

## EDWARD PAGE MITCHELL

—

# The Tachypomp

$\times \div \times$
$+$

THERE WAS NOTHING mysterious about Professor Surd's dislike for me. I was the only poor mathematician in an exceptionally mathematical class. The old gentleman sought the lecture room every morning with eagerness, and left reluctantly. For was it not a thing of joy to find seventy young men who, individually and collectively, preferred x to XX; who had rather differentiate than dissipate; and for whom the limbs of the heavenly bodies had more attractions than those of earthly stars upon the spectacular stage?

So affairs went on swimmingly between the professor of mathematics and the junior class at Polyp University. In every man of the seventy the sage saw the logarithm of a possible La Place, of a Sturm, or of a Newton. It was a delightful task for him to lead them through the pleasant valleys of conic sections, and beside the still waters of the integral calculus. Figuratively speaking, his problem was not a hard one. He had only to manipulate, and eliminate, and to raise to a higher power, and the triumphant result of examination day was assured.

But I was a disturbing element, a perplexing unknown quantity, which had somehow crept into the work, and which seriously threatened to impair the accuracy of his calculations. It was a touching sight to behold the venerable mathematician as he pleaded with me not so utterly to disregard precedent in the use of cotangents; or as he urged, with eyes almost tearful, that ordinates were dangerous things to trifle with. All in vain. More theorems went onto my cuff than into my head. Never did chalk do so much work to so little purpose. And, therefore, it came that Furnace Second was reduced to zero in Professor Surd's estimation. He looked upon me with all the horror which an

[FIRST PUBLISHED IN SCRIBNER'S MAGAZINE IN 1873]

unalgebraic nature could inspire. I have seen the professor walk around the entire square rather than meet the man who had no mathematics in his soul.

For Furnace Second were no invitations to Professor Surd's house. Seventy of the class supped in delegations around the periphery of the professor's tea-table. The seventy-first knew nothing of the charms of that perfect ellipse, with its twin bunches of fuchsias and geraniums in gorgeous precision at the two foci.

This, unfortunately enough, was no trifling deprivation. Not that I longed especially for segments of Mrs. Surd's justly celebrated lemon pies; not that the spheroidal damsons of her excellent preserving had any marked allurements; not even that I yearned to hear the professor's jocose table talk about binomials, and chatty illustrations of abstruse paradoxes. The explanation is far different. Professor Surd had a daughter. Twenty years before, he made a proposition of marriage to the present Mrs. S. He added a little corollary to his proposition not long after. The corollary was a girl.

Abscissa Surd was as perfectly symmetrical as Giotto's circle, and as pure, withal, as the mathematics her father taught. It was just when spring was coming to extract the roots of frozen-up vegetation that I fell in love with the corollary. That she herself was not indifferent I soon had reason to regard as a self-evident truth.

The sagacious reader will already recognize nearly all the elements necessary to a well-ordered plot. We have introduced a heroine, inferred a hero, and constructed a hostile parent after the most approved model. A movement for the story, a *deus ex machina*, is alone lacking. With considerable satisfaction I can promise a perfect novelty in this line, a *deus ex machina* never before offered to the public.

It would be discounting ordinary intelligence to say that I sought with unwearying assiduity to figure my way into the stern father's good will; that never did dullard apply himself to mathematics more patiently than I; that never did faithfulness achieve such meager reward. Then I engaged a private tutor. His instructions met with no better success.

My tutor's name was Jean Marie Rivarol. He was a unique Alsatian —though Gallic in name, thoroughly Teuton in nature; by birth, a Frenchman; by education, a German. His age was thirty; his profession, omniscience; the wolf at his door, poverty; the skeleton in his closet, a consuming but unrequited passion. The most recondite principles of practical science were his toys; the deepest intricacies of abstract science his diversions. Problems which were foreordained mysteries to me were to him as clear as Tahoe water. Perhaps this very fact will explain our lack of success in the relation of tutor and pupil; perhaps the failure is alone due to my own unmitigated stupidity. Rivarol had hung about the skirts of the University for several years, supplying his few wants by writing for scientific journals, or by giving assistance to students who, like myself, were characterized by a plethora of purse and a paucity of ideas; cooking, studying, and sleeping in his attic lodgings; and prosecuting queer experiments all by himself.

We were not long discovering that even this eccentric genius could not transplant brains into my deficient skull. I gave over the struggle in despair. An unhappy year dragged its slow length around. A gloomy year it was, brightened only by occasional interviews with Abscissa, the Abbie of my thoughts and dreams.

Commencement day was coming on apace. I was soon to go forth, with the rest of my class, to astonish and delight a waiting world. The professor seemed to avoid me more than ever. Nothing but the conventionalities, I think, kept him from shaping his treatment of me on the basis of unconcealed disgust.

At last, in the very recklessness of despair, I resolved to see him, plead with him, threaten him if need be, and risk all my fortunes on one desperate chance. I wrote him a somewhat defiant letter, stating my aspirations and, as I flattered myself, shrewdly giving him a week to get over the first shock of horrified surprise. Then I was to call and learn my fate.

During the week of suspense I nearly worried myself into a fever. It was first crazy hope, and then saner despair. On Friday evening, when I presented myself at the professor's door, I was such a haggard, sleepy, dragged-out specter that even Miss Jocasta, the harsh-favored maiden sister of the Surds, admitted me with commiserate regard, and suggested pennyroyal tea.

Professor Surd was at a faculty meeting. Would I wait?

Yes, till all was blue, if need be. Miss Abbie?

Abscissa had gone to Wheelborough to visit a school friend. The aged maiden hoped I would make myself comfortable, and departed to the unknown haunts which knew Jocasta's daily walk.

Comfortable! But I settled myself in a great uneasy chair and waited, with the contradictory spirit common to such junctures, dread-ing every step lest it should herald the man whom, of all men, I wished to see.

I had been there at least an hour, and was growing right drowsy.

At length Professor Surd came in. He sat down in the dusk opposite me, and I thought his eyes glinted with malignant pleasure as he said, abruptly, "So, young man, you think you are a fit husband for my girl?"

I stammered some inanity about making up in affection what I lacked in merit; about my expectations, family, and the like. He quickly interrupted me.

"You misapprehend me, sir. Your nature is destitute of those mathematical perceptions and acquirements which are the only sure foundations of character. You have no mathematics in you. You are fit for treasons, stratagems, and spoils.—Shakespeare. Your narrow intellect cannot understand and appreciate a generous mind. There is all the difference between you and a Surd, if I may say it, which intervenes between an infinitesimal and an infinite. Why, I will even venture to say that you do not comprehend the Problem of the Couriers!"

I admitted that the Problem of the Couriers should be classed rather without my list of accomplishments than within it. I regretted this fault very deeply, and suggested amendment. I faintly hoped that my fortune would be such—

"Money!" he impatiently exclaimed. "Do you seek to bribe a Roman senator with a penny whistle? Why, boy, do you parade your paltry wealth, which, expressed in mills, will not cover ten decimal places, before the eyes of a man who measures the planets in their orbits, and close crowds infinity itself?"

I hastily disclaimed any intention of obtruding my foolish dollars, and he went on:

"Your letter surprised me not a little. I thought you would be the last person in the world to presume to an alliance here. But having a regard for you personally—" and again I saw malice twinkle in his small eyes—"and still more regard for Abscissa's happiness, I have decided that you shall have her—upon conditions. Upon conditions," he repeated, with a half-smothered sneer.

"What are they?" cried I, eagerly enough. "Only name them."

"Well, sir," he continued, and the deliberation of his speech seemed the very refinement of cruelty, "you have only to prove yourself worthy an alliance with a mathematical family. You have only to accomplish a task which I shall presently give you. Your eyes ask me what it is. I will tell you. Distinguish yourself in that noble branch of abstract science in which, you cannot but acknowledge, you are at present sadly deficient. I will place Abscissa's hand in yours whenever you shall come before me and square the circle to my satisfaction. No! That is too easy a condition. I should cheat myself. Say perpetual motion. How do you like that? Do you think it lies within the range of your mental capabilities? You don't smile. Perhaps your talents don't run in the way of perpetual motion. Several people have found that theirs didn't. I'll give you another chance. We were speaking of the Problem of the Couriers, and I think you expressed a desire to know more of that ingenious question. You shall have the opportunity. Sit down some day, when you have nothing else to do, and discover the principle of infinite speed. I mean the law of motion which shall accomplish an infinitely great distance in an infinitely short time. You may mix in a little practical mechanics, if you choose. Invent some method of taking the tardy Courier over his road at the rate of sixty miles a minute. Demonstrate me this discovery (when you have made it!) mathematically, and approximate it practically, and Abscissa is yours. Until you can, I will thank you to trouble neither myself nor her."

I could stand his mocking no longer. I stumbled mechanically out of the room, and out of the house. I even forgot my hat and gloves. For an hour I walked in the moonlight. Gradually I succeeded to a more hopeful frame of mind. This was due to my ignorance of mathematics. Had I understood the real meaning of what he asked, I should have been utterly despondent.

Perhaps this problem of sixty miles a minute was not so impossible after all. At any rate I could attempt, though I might not succeed. And Rivarol came to my mind. I would ask him. I would enlist his knowledge to accompany my own devoted perseverance. I sought his lodgings at once.

The man of science lived in the fourth story back. I had never been in his room before. When I entered, he was in the act of filling a beer mug from a carboy labeled *aqua fortis*.

"Seat you," he said. "No, not in that chair. That is my petty-cash adjuster." But he was a second too late. I had carelessly thrown myself into a chair of seductive appearance. To my utter amazement it reached out two skeleton arms and clutched me with a grasp against which I struggled in vain. Then a skull stretched itself over my shoulder and grinned with ghastly familiarity close to my face.

Rivarol came to my aid with many apologies. He touched a spring somewhere and the petty-cash adjuster relaxed its horrid hold. I placed myself in a plain cane-bottomed rocking-chair, which Rivarol assured me was a safe location.

"That seat," he said, "is an arrangement upon which I much felicitate myself. I made it at Heidelberg. It has saved me a vast deal of small annoyance. I consign to its embraces the friends who bore, and the visitors who exasperate, me. But it is never so useful as when terrifying some tradesman with an insignificant account. Hence the pet name which I have facetiously given it. They are invariably too glad to purchase release at the price of a bill receipted. Do you well apprehend the idea?"

While the Alsatian diluted his glass of *aqua fortis*, shook into it an infusion of bitters, and tossed off the bumper with apparent relish, I had time to look around the strange apartment.

The four corners of the room were occupied respectively by a turning-lathe, a Ruhmkorff coil, a small steam-engine, and an orrery in stately motion. Tables, shelves, chairs, and floor supported an odd aggregation of tools, retorts, chemicals, gas-receivers, philosophical instruments, boots, flasks, paper-collar boxes, books diminutive, and books of preposterous size. There were plaster busts of Aristotle, Archimedes, and Compte, while a great drowsy owl was blinking away, perched on the benign brow of Martin Farquhar Tupper. "He always roosts there when he proposes to slumber," explained my tutor. "You are a bird of no ordinary mind. *Schlafen Sie wohl.*"

Through a closet door, half open, I could see a humanlike form covered with a sheet. Rivarol caught my glance.

"That," said he, "will be my masterpiece. It is a microcosm, an android, as yet only partially complete. And why not? Albertus Magnus constructed an image perfect to talk metaphysics and confute the schools. So did Sylvester II; so did Robertus Greathead. Roger Bacon made a brazen head that held discourses. But the first named of these came to destruction. Thomas Aquinas got wrathful at some of its syllogisms and smashed its head. The idea is reasonable enough. Mental action will yet be reduced to laws as definite as those which govern the physical. Why should not I accomplish a manikin which shall preach as original discourses as the Rev. Dr. Allchin, or talk poetry as mechanically as Paul Anapest? My android can already work problems in vulgar fractions and compose sonnets. I hope to teach it the positive philosophy."

Out of the bewildering confusion of his effects Rivarol produced two pipes and filled them. He handed one to me.

"And here," he said, "I live and am tolerably comfortable. When my coat wears out at the elbows I seek the tailor and am measured for another. When I am hungry I promenade myself to the butcher's and bring home a pound or so of steak, which I cook very nicely in three seconds by this oxyhydrogen flame. Thirsty, perhaps, I send for a carboy of *aqua fortis*. But I have it charged, all charged. My spirit is above any small pecuniary transaction. I loathe your dirty greenbacks and never handle what they call scrip."

"But are you never pestered with bills?" I asked. "Don't the creditors worry your life out?"

"Creditors!" gasped Rivarol. "I have learned no such word in your very admirable language. He who will allow his soul to be vexed by creditors is a relic of an imperfect civilization. Of what use is science if it cannot avail a man who has accounts current? Listen. The moment you or any one else enters the outside door this little electric bell sounds me warning. Every successive step on Mrs. Grimler's staircase is a spy and informer vigilant for my benefit. The first step is trod upon. That trusty first step immediately telegraphs your weight. Nothing could be simpler. It is exactly like any platform scale. The weight is registered up here upon this dial. The second step records the size of my visitor's feet. The third his height, the fourth his complexion, and so on. By the time he reaches the top of the first flight I have a pretty accurate description of him right here at my elbow, and quite a margin of time for deliberation and action. Do you follow me? It is plain enough. Only the A B C of my science."

"I see all that," I said, "but I don't see how it helps you any. The knowledge that a creditor is coming won't pay his bill. You can't escape unless you jump out of the window."

Rivarol laughed softly. "I will tell you. You shall see what becomes of any poor devil who goes to demand money of me—of a man of science. Ha! ha! It pleases me. I was seven weeks perfecting my dun suppressor. Did you know—" he whispered exultingly—"did you know that there is a hole through the earth's center? Physicists have long suspected it; I was the first to find it. You have read how Rhuyghens, the Dutch navigator, discovered in Kerguelen's Land an abysmal pit which fourteen hundred fathoms of plumb-line failed to sound. Herr Tom, that hole has no bottom! It runs from one surface of the earth to the antipodal surface. It is diametric. But where is the antipodal spot? You stand upon it. I learned this by the merest chance. I was deep-digging in Mrs. Grimler's cellar, to bury a poor cat I had sacrificed in a galvanic experiment, when the earth under my spade crumbled, caved in, and wonder-stricken I stood upon the brink of a yawning shaft. I dropped a coal-hod in. It went down, down, down, bounding and rebounding. In two hours and a quarter that coal-hod came up again. I caught it and restored it to the angry Grimler. Just think a minute. The coal-hod went down, faster and faster, till it reached the center of the earth. There it would stop, were it not for acquired momentum. Beyond the center its journey was relatively upward, toward the opposite surface of the globe. So, losing velocity, it went slower and slower till it reached that surface. Here it came to rest for a second and then fell back again, eight thousand odd miles, into my hands. Had I not interfered with it, it would have repeated its journey, time after time, each trip of shorter extent, like the diminishing oscillations of a pendulum, till it finally came to eternal rest at the center of the sphere. I am not slow to give a practical application to any such grand discovery. My dun suppressor was born of it. A trap, just outside my chamber door; a spring in her; a creditor on the trap: need I say more?"

"But isn't it a trifle inhuman?" I mildly suggested. "Plunging an unhappy being into a perpetual journey to and from Kerguelen's Land, without a moment's warning."

"I give them a chance. When they come up the first time I wait at the mouth of the shaft with a rope in hand. If they are reasonable and will come to terms, I fling them the line. If they perish, 'tis their own fault. Only," he added, with a melancholy smile, "the center is getting so plugged up with creditors that I am afraid there soon will be no choice whatever for 'em."

By this time I had conceived a high opinion of my tutor's ability. If anybody could send me waltzing through space at an infinite speed, Rivarol could do it. I filled my pipe and told him the story. He heard with grave and patient attention. Then, for full half an hour, he whiffed away in silence. Finally he spoke:

"The ancient cipher has overreached himself. He has given you a choice of two problems, both of which he deems insoluble. Neither of them is insoluble. The only gleam of intelligence Old Cotangent showed was when he said that squaring the circle was too easy. He was right. It would have given you your *Liebchen* in five minutes. I squared the circle before I discarded pantalets. I will show you the work—but it would be a digression, and you are in no mood for digressions. Our first chance, therefore, lies in perpetual motion. Now, my good friend, I will frankly tell you that, although I have compassed this interesting problem, I do not choose to use it in your behalf. I too, Herr Tom, have a heart. The loveliest of her sex frowns upon me. Her somewhat mature charms are not for Jean Marie Rivarol. She has cruelly said that her years demand of me filial rather than connubial regard. Is love a matter of years or of eternity? This question did I put to the cold, yet lovely, Jocasta."

"Jocasta Surd!" I remarked in surprise. "Abscissa's aunt!"

"The same," he said sadly. "I will not attempt to conceal that upon the maiden Jocasta my maiden heart has been bestowed. Give me your hand, my nephew in affliction as in affection!"

Rivarol dashed away a not discreditable tear, and resumed:

"My only hope lies in this discovery of perpetual motion. It will give me the fame, the wealth. Can Jocasta refuse these? If she can, there is only the trap door and—Kerguelen's Land!"

I bashfully asked to see the perpetual-motion machine. My uncle in affliction shook his head.

"At another time," he said. "Suffice it at present to say that it is something upon the principle of a woman's tongue. But you see now why we must turn in your case to the alternative condition, infinite speed. There are several ways in which this may be accomplished, theoretically. By the lever, for instance. Imagine a lever with a very long and a very short arm. Apply power to the shorter arm which will move it with great velocity. The end of the long arm will move much faster. Now keep shortening the short arm and lengthening the long one, and as you approach infinity in their difference of length, you approach infinity in the speed of the long arm. It would be difficult to demonstrate this practically to the professor. We must seek another solution. Jean Marie will meditate. Come to me in a fortnight. Good night. But stop! Have you the money—*das Geld?*"

"Much more than I need."

"Good! Let us strike hands. Gold and Knowledge; Science and Love. What may not such a partnership achieve? We go to conquer thee, Abscissa. *Vorwärts!*"

When, at the end of a fortnight, I sought Rivarol's chamber, I passed with some little trepidation over the terminus of the air line to Kerguelen's Land, and evaded the extended arms of the petty-cash adjuster. Rivarol drew a mug of ale for me, and filled himself a retort of his own peculiar beverage.

"Come," he said at length. "Let us drink success to the Tachypomp."

"The Tachypomp?"

"Yes. Why not? *Tachu,* quickly, and *pempo, pepompa,* to send. May it send you quickly to your wedding day. Abscissa is yours. It is done. When shall we start for the prairies?"

"Where is it?" I asked, looking in vain around the room for any contrivance which might seem calculated to advance matrimonial prospects.

"It is here." And he gave his forehead a significant tap. Then he held forth didactically.

"There is force enough in existence to yield us a speed of sixty miles a minute, or even more. All we need is the knowledge how to combine and apply it. The wise man will not attempt to make some great force yield some great speed. He will keep adding the little force to the little force, making each little force yield its little speed, until an aggregate of little forces shall be a great force, yielding an aggregate of little speeds, a great speed. The difficulty is not in aggregating the forces; it lies in the corresponding aggregation of the speeds. One musket-ball will go, say a mile. It is not hard to increase the force of muskets to a thousand, yet the thousand musket balls will go no farther, and no faster, than the one. You see, then, where our trouble lies. We cannot readily add speed to speed, as we add force to force. My discovery is simply the utilization of a principle which extorts an increment of speed from each increment of power. But this is the metaphysics of physics. Let us be practical or nothing.

"When you have walked forward, on a moving train, from the rear car, toward the engine, did you ever think what you were really doing?"

"Why, yes, I have generally been going to the smoking car to have a cigar."

"Tut, tut—not that! I mean did it ever occur to you on such an occasion that absolutely you were moving faster than the train? The train passes the telegraph poles at the rate of thirty miles an hour, say. You walk towards the smoking car at the rate of four miles an hour.

Then you pass the telegraph poles at the rate of thirty-four miles. Your absolute speed is the speed of the engine, plus the speed of your own locomotion. Do you follow me?"

I began to get an inkling of his meaning, and told him so.

"Very well. Let us advance a step. Your addition to the speed of the engine is trivial, and the space in which you can exercise it limited. Now suppose two stations, A and B, two miles distant by the track. Imagine a train of platform cars, the last car resting at station A. The train is a mile long, say. The engine is therefore within a mile of station B. Say the train can move a mile in ten minutes. The last car, having two miles to go, would reach B in twenty minutes, but the engine, a mile ahead, would get there in ten. You jump on the last car, at A, in a prodigious hurry to reach Abscissa, who is at B. If you stay on the last car it will be twenty long minutes before you see her. But the engine reaches B and the fair lady in ten. You will be a stupid reasoner, and an indifferent lover, if you don't put for the engine over those platform cars, as fast as your legs will carry you. You can run a mile, the length of the train, in ten minutes. Therefore, you reach Abscissa when the engine does, or in ten minutes—ten minutes sooner than if you had lazily sat down upon the rear car, and talked politics with the brakeman. You have diminished the time by one-half. You have added your speed to that of the locomotive to some purpose. *Nicht wahr?*"

I saw it perfectly; much plainer, perhaps, for his putting in the clause about Abscissa.

He continued:

"This illustration, though a slow one, leads up to a principle which may be carried to any extent. Our first anxiety will be to spare your legs and wind. Let us suppose that the two miles of track are perfectly straight, and make our train one platform car, a mile long, with parallel rails laid upon its top. Put a little dummy engine on these rails, and let it run to and fro along the platform car, while the platform car is pulled along the ground track. Catch the idea? The dummy takes your place. But it can run its mile much faster. Fancy that our locomotive is strong enough to pull the platform car over the two miles in two minutes. The dummy can attain the same speed. When the engine reaches B in one minute, the dummy, having gone a mile atop the platform car, reaches B also. We have so combined the speeds of those two engines as to accomplish two miles in one minute. Is this all we can do? Prepare to exercise your imagination."

I lit my pipe.

"Still two miles of straight track, between A and B. On the track a long platform car, reaching from A to within a quarter of a mile of B. We will now discard ordinary locomotives and adopt as our motive power a series of compact magnetic engines, distributed underneath the platform car, all along its length."

"I don't understand those magnetic engines."

"Well, each of them consists of a great iron horseshoe, rendered alternately a magnet and not a magnet by an intermittent current of electricity from a battery, this current in its turn regulated by clockwork. When the horseshoe is in the circuit it is a magnet, and it pulls its clapper toward it with enormous power. When it is out of the circuit, the next second, it is not a magnet and it lets the clapper go. The clapper, oscillating to and fro, imparts a rotatory motion to a flywheel, which transmits it to the drivers on the rails. Such are our motors. They are no novelty, for trial has proved them practicable.

"With a magnetic engine for every truck of wheels, we can reasonably expect to move our immense car, and to drive it along at a speed, say, of a mile a minute.

"The forward end, having but a quarter of a mile to go, will reach B in fifteen seconds. We will call this platform car number 1. On top of number 1 are laid rails on which another platform car, number 2, a quarter of a mile shorter than number 1, is moved in precisely the same way. Number 2, in its turn, is surmounted by number 3, moving independently of the tiers beneath, and a quarter of a mile shorter than number 2. Number 2 is a mile and a half long; number 3, a mile and a quarter. Above, on successive levels, are number 4, a mile long; number 5, three-quarters of a mile; number 6, half a mile; number 7, a quarter of a mile, and number 8, a short passenger car, on top of all.

"Each car moves upon the car beneath it, independently of all the others, at the rate of a mile a minute. Each car has its own magnetic engines. Well, the train being drawn up with the latter end of each car resting against a lofty bumping-post at A, Tom Furnace, the gentlemanly conductor, and Jean Marie Rivarol, engineer, mount by a long ladder to the exalted number 8. The complicated mechanism is set in motion. What happens?

"Number 8 runs a quarter of a mile in fifteen seconds and reaches the end of number 7. Meanwhile number 7 has run a quarter of a mile in the same time and reached the end of number 6; number 6, a quarter of a mile in fifteen seconds, and reached the end of number 5; number 5, the end of number 4; number 4, of number 3; number 3, of number 2; number 2, of number 1. And number 1, in fifteen seconds, has gone its quarter of a mile along the ground track, and has reached station B. All this has been done in fifteen seconds. Wherefore numbers 1, 2, 3, 4, 5, 6, 7, and 8 come to rest against the bumping-post at B at precisely the same second. We, in number 8, reach B just when number 1 reaches it. In other words, we accomplish two miles in fifteen seconds. Each of the eight cars, moving at the rate of a mile a minute, has contributed a quarter of a mile to our journey, and has done its work in fifteen seconds. All the eight did their work at once, during the same fifteen seconds. Consequently we have been whizzed through the air at the somewhat startling speed of seven and a half seconds to the mile. This is the Tachypomp. Does it justify the name?"

Although a little bewildered by the complexity of cars, I apprehended the general principle of the machine. I made a diagram and understood it much better. "You have merely improved on the idea of my moving faster than the train when I was going to the smoking car?"

"Precisely. So far we have kept within the bounds of the practicable. To satisfy the professor you can theorize in something after this fashion: If we double the number of cars, thus decreasing by one half the distance which each has to go, we shall attain twice the speed. Each of the sixteen cars will have but one eighth of a mile to go. At the uniform rate we have adopted, the two miles can be done in seven and a half instead of fifteen seconds. With thirty-two cars, and a sixteenth of a mile, or twenty rods difference in their length, we arrive at the speed of a mile in less than two seconds; with sixty-four cars, each traveling but ten rods, a mile under the second. More than sixty miles a minute! If this isn't rapid enough for the professor, tell him to go on, increasing the number of his cars and diminishing the distance each one has to run. If sixty-four cars yield a speed of a mile inside the second, let him fancy a Tachypomp of six hundred and forty cars, and amuse himself calculating the rate of car number 640. Just whisper to him that when he has an infinite number of cars with an infinitesimal difference in their lengths, he will have obtained that infinite speed for which he seems to yearn. Then demand Abscissa."

I wrung my friend's hand in silent and grateful admiration. I could say nothing.

"You have listened to the man of theory," he said proudly. "You shall now behold the practical engineer. We will go to the west of the Mississippi and find some suitably level locality. We will erect thereon a Model Tachypomp. We will summon thereunto the professor, his daughter, and why not his fair sister Jocasta, as well? We will take them a journey which shall much astonish the venerable Surd. He shall place Abscissa's digits in yours and bless you both with an algebraic formula. Jocasta shall contemplate with wonder the geinus of Rivarol. But we have much to do. We must ship to St. Joseph the vast amount of material to be employed in the construction of the Tachypomp. We must engage a small army of workmen to effect that construction, for we are to annihilate time and space. Perhaps you had better see your bankers."

I rushed impetuously to the door. There should be no delay.

"Stop! stop! *Um Gottes Willen*, stop!" shrieked Rivarol. "I launched my butcher this morning and I haven't bolted the—"

But it was too late. I was up on the trap. It swung open with a crash and I was plunged down, down, down! I felt as if I were falling through illimitable space. I remember wondering, as I rushed through the darkness, whether I should reach Kerguelen's Land, or stop at the center. It seemed an eternity. Then my course was suddenly and painfully arrested.

I opened my eyes. Around me were the walls of Professor Surd's study. Under me was a hard, unyielding plane which I knew too well was Professor Surd's study floor. Behind me was the black, slippery haircloth chair which had belched me forth, much as the whale served Jonah. In front of me stood Professor Surd himself, looking down with a not unpleasant smile.

"Good evening, Mr. Furnace. Let me help you up. You look tired, sir. No wonder you fell asleep when I kept you so long waiting. Shall I get you a glass of wine? No? By the way, since receiving your letter I find that you are a son of my old friend, Judge Furnace. I have made inquiries, and see no reason why you should not make Abscissa a good husband."

Still I can see no reason why the Tachypomp should not have succeeded. Can you?

MARTIN GARDNER

—

# The Island of Five Colors

× ÷ ×
+

THERE IS ONLY one paint store in Monrovia, the capital of Liberia. When I told the ebony-skinned clerk how many gallons of paint I wanted, he lifted his bushy eyebrows and whistled through his teeth.

"You musta be goin' to paint a whole mountain, mistah!"

I said, "No—just an island."

He grinned. He thought I was joking. But I really *was* planning to paint an island. I planned to paint it five colors—red, blue, green, yellow, and purple.

Why? To answer that question I'll have to go back several years and explain my interest in the "four-color theorem"—the most famous unsolved problem in topology. It was in 1947 that Professor Stanislaw Slapenarski, of the University of Warsaw, gave a series of lectures at the University of Chicago on topology and relativity. I was at that time an instructor (I have since been advanced to assistant professor) in the university's math department. Slapenarski and I became good friends, and I had the privilege of introducing him to the Moebius Society the evening he gave his sensational lecture on the "no-sided surface."[1] Readers familiar with Slapenarski's career will recall his unfortunate heart attack and death early in 1948, after his return to Warsaw.

The four-color theorem had been the subject of my Ph.D. thesis, and a topic on which Slapenarski and I exchanged many letters before his visit to the States. The theorem asserts that four colors are all one needs for coloring any conceivable type of map in such a way that no

[1] I have told the story of Slapenarski's discovery of non-lateral surfaces, followed by his untimely heart attack, in "No-Sided Professor." See page 99.

two bordering countries have the same color (countries which touch at a single point are not considered "bordering"). The districts may be any size or shape, however curious, and there may be any number of them. The theorem was first mentioned in 1860 by Moebius, one of the great pioneers of topology, and although the finest minds in mathematics have worked on it since, no formal proof for it has yet been found.[2]

Oddly enough, the problem has been solved for other surfaces than a plane.[3] In 1890 P. J. Heawood proved that *seven* colors were necessary and sufficient for a torus (doughnut-shaped) surface, and in 1934 Philip Franklin proved that *six* colors were necessary and sufficient for maps on one-sided surfaces such as the Moebius strip and the Klein bottle.[4]

Slapenarski's discovery of the no-sided surface had important bearings on the properties of a Klein bottle, and revolutionary implications for the study of the four-color theorem. I have a vivid memory of the rotund Polish professor smiling and pulling on his reddish beard, and saying, "My dear Martin, if the history of topology teaches anything, it is that one must expect the most surprising connections between apparently unrelated topological problems."

It was in line with some of Slapenarski's suggestions that I wrote in 1950 my historic refutation of a previous "proof" by Heawood that *five* colors were sufficient for maps on a plane surface. Topologists remained convinced, of course, that the four-color theorem was true, but in the light of these new developments, the formal "proof" of the theorem seemed farther away than ever.

Shortly after I published the paper mentioned above, I had lunch at the university's Quadrangle Club with Dr. Alma Bush. Alma is one of the nation's leading cultural anthropologists, and certainly the

[2] In 1879 A. B. Kempe published a "proof" of the four-color theorem, but ten years later P. J. Heawood found a mistake in it.

[3] The problem for the surface of a globe is the same as for a plane, since the globe can be punctured in the middle of any region and flattened out.

[4] The true Klein bottle, named after Felix Klein, the German topologist who edited the works of Moebius, is a closed surface (like that of a balloon), but due to its one-sided character, it has neither outside nor inside. A twist through the fourth dimension is necessary for a construction of the bottle, which does not intersect itself. The familiar model is not a true Klein bottle because of the puncturing which forms a hole. For a full discussion of the bottle's properties see p. 271f of *Anschauliche Geometrie*, by D. Hillbert and S. Cohn-Vossen, Berlin, 1932.

handsomest woman on the college faculty, She was in her late thirties, but her figure was remarkably youthful, and the curves under the lapels of her tweed jacket were anything but masculine. Her eyes were pale gray, and she had a habit of squinting them slightly when she was thinking.

Alma had just returned from a small island a few hundred miles off the coast of Liberia, on the west side of the African continent. She was in charge of a group of anthropology students who were studying the five tribes that inhabited the island. Apparently the tribes are of great interest to anthropologists because of the astonishing variations in customs among them.

"The island's divided into five districts," Alma said, putting a cigarette in her long, black plastic holder. "They all border on each other. It's important to an understanding of their *mores.* The fact that all the tribes connect with common borders helps weld them into a kind of cultural unity. Why, Marty! Why are you looking so surprised?"

I had stopped eating with a fork in mid-air. I put it down slowly. "Because," I said, "what you just told me is impossible."

Alma looked irritated. "What's impossible?"

"That the five tribes have common borders. It violates the famous four-color theorem."

"The what?"

"The four-color map theorem," I said. "It's a theorem in topology." I drew diagrams on the tablecloth with the end of my spoon and tried to explain it to her.

Alma seemed to get the general idea. "Maybe these tribes use a different kind of mathematics?" she suggested, squinting at me through her cigarette smoke.

I shook my head. "Mathematics, my dear, is the same in all cultures. Two plus two is never anything but four—even in Africa."[5]

But Alma didn't think so. She said there were "great cultural variations" in the mathematical thinking of primitive societies, and she even knew of one obscure tribe that believed whenever two canoes were added to two canoes, the result was *five* canoes.

"Then they were in error," I said.

"From your point of view," Alma added, her gray eyes smiling.

"Look, Alma," I said, between mouthfuls of raspberry sherbet, "if

[5] See my paper on "Mathematics and the Folkways," *Journal of Philosophy*, March 30, 1950.

your island is really divided as you say, with each of the five districts bordering on the other four, then I'll admit mathematics is in the folkways. Do you have a map of this island?"

She shook her head. "It's never been mapped. I hope we'll be able to map it when I get back."

Naturally, I didn't believe her. But she stuck to the story, and for the life of me I couldn't tell whether she was actually convinced it was true, or whether she was trying to pull my leg.

"Why don't you come along and see for yourself?" she said, with a flourish of her cigarette holder. "I'll only be there a month or so—just long enough to recheck some of my data before it's published. You can help map the island. If what I said isn't true, I'll pay your expenses."

What could I lose? I suspected, of course, a joker somewhere; but I had a vacation coming up, and it would be a pleasant, unusual trip. I'd always wanted to see how anthropologists went about their field work. Besides, when Alma and I aren't arguing, we get along fine.

We took a steamer from New York to Monrovia. There's a small airfield there, so we were able to get to the island by plane. The plane made weekly trips back and forth, carrying supplies to the island and bringing back palm cones (a source of palm oil) and coffee beans, the island's two chief exports. Alma assigned me to a tent in the camp where the anthropology students had set up headquarters.

The sky was cloudless and the sun beat down with an oppressive, prickly heat. I had outfitted myself with a pair of khaki shorts, a khaki shirt, and a big tropical pith hclmct. Thc hclmct protected the bald part of my head from thc blazing sun. Alma also wore shorts, and I must confess she looked much better in them than I did. My legs are rather long and bony.

The camp was pitched on a small clearing near the edge of a dense growth of tangled brush and luxuriant, waving ferns. When you walked about, hundreds of small lizards seemed to scurry out of your path. There was a constant buzzing of flies and mosquitoes, but we plastered the exposed parts of our anatomy with foul-smelling repellent, and the insects bothered us less than I expected.

On the second day after our arrival Alma introduced me to Aguz, one of her new contacts on the island. Aguz could not speak English, but Alma had mastered the native dialect well enough to converse with him. He was a tall, good-looking black, with high cheekbones, brilliant white teeth, and deep purplish-brown skin. He was bare to the waist except for a small bow tie around his neck. His trousers were several inches above the ankles, exposing brightly-checked red and yellow socks. There were no shoes, and I noticed that the soles of the socks had worn off, so actually he padded around barefooted. A Phi Beta Kappa key hung from a leather belt. When we shook hands he grunted something that Alma translated as meaning he was delighted to meet me.

Aguz was one of the members of the Hiyiku tribe, the intellectual elite of the island. Ten years ago, Alma explained, a group of Princeton anthropologists had studied the tribe. The Hiyikus had adopted their style of dress.

Alma arranged with Aguz for the three of us to make a walking tour of the island. Fortunately, it was a small island, not more than twenty-five square miles in area, and by starting early in the morning, we could make the tour easily in a day. I took along a pad of paper and a box of crayons so I could make a rough map of the five districts.

Our first visit was to the Hiyikus, on whose territory the camp was pitched. Their village was an area of neat, circular mud huts with dirt floors and cone-shaped roofs of thatched grass. The tribesmen dressed like Aguz, except for the gold key which Alma said had been given him by one of the Princeton professors. They stood outside their huts, smoking pipes and watching us with grave, philosophic expressions. The women were nude above the waist, with bright dyed breechcloths wrapped around their hips. They wore harlequin glasses and large brass rings in their noses. Most of them were sitting in little groups weaving mats out of palm fiber.

I made a circle on the top sheet of my tablet and colored it blue. I didn't know the exact shape of the Hiyiku area, but this would do well enough for my purposes. When we crossed westward into the Wolfezi territory, I added a green patch to the left of the blue.

I picked green because the Wolfezis wore nothing but long strings of green beads. The tribe consisted entirely of males. Alma explained their cultural organization in some detail, but all I need say here is that the men were bachelors recruited from the other four tribes. They formed a kind of reservoir of men on which the other tribes drew whenever sickness or warfare reduced their male ratio.

The Wolfezis sang and danced a great deal, and impressed me as being a happy, contented group. Their life was marred, Alma explained, only by constant fear of being chosen by lot to become husbands in one of the other tribes. Frequently the unlucky ones would kill themselves by jumping off a large cliff. Aguz pointed it out to me. It was known locally as the "kala ulukiffir" or "cliff of the unfortunates."

The chief broke a path through some dense underbrush, leading us down an embankment to a narrow, sluggish river. A large raft of logs was half buried in the thick yellow mud that formed the river bed. Aguz shoved the raft into the water, climbed aboard, then steadied it with a long bamboo pole while we slogged through the mud and joined him. Using the pole, he pushed the raft slowly down the twisting river.

Giant palm trees arched over us on both sides, blocking out the sunlight except for occasional rays that broke through the foliage and made shimmering patterns on the brown surface of the river. Now and then we passed enormous, mud-caked crocodiles, motionless in the shallow water.

The chief's bow tie bobbed up and down as he made guttural sounds which Alma translated as meaning we were entering the Gesellomo region. It was north of the Wolfezi district. I brushed a huge dragonfly off my tablet and added a red patch to the map, just above the green.

After moving a half-mile or so into Gesellomo territory, Aguz banked the raft and we sloshed our way to dry land. A short climb up the slope, through tall, thick grass, and we found ourselves at the edge of the Gesellomo village.

I have no intention of describing in detail this remarkable tribe, because Dr. Bush has covered it so thoroughly in her forthcoming book on the island. I'll limit myself, therefore, to pointing out that the Gesellomo culture is the only known primitive society organized on what Alma called the "filiarchal" basis. From the age of one to five, children are under the control of parents. After that, the children take over, and local mores bind the parents to absolute obedience.

We were careful to skirt the village because, as Aguz warned, the children assume all adults to be of a class with their own, and so would attempt to put us to work. I could see the older folks performing various menial tasks while the children stood around watching and playing. One boy of about seven was thrashing his father vigorously with a whip of leather thongs.

We tramped off to the southeast. After going about a mile, Aguz pointed to a row of palm trees ahead and said they marked the boundary between the Gesellomos and the Hiyikus. I got out my tablet and expanded the red area until it bordered on the blue.

The difficulty of hiking through dense jungle underbrush, and the unrelenting glare of the tropical sun, made us hot and tired and hungry. We rested on some large rocks of red sandstone and ate our lunch. In the distance I could hear the faint beating of drums.

It was early afternoon when we reached the village of the Bebopulus. The males of this curious tribe wore dark shell-rimmed glasses and had tiny goatees on their chins. Both sexes had the odd practice of inserting a large bone through their heads so that it projected from each ear. The custom not only rendered them totally deaf, but also damaged their brains. They communicated with each other by simple signs.

We passed a group of them, wearing loincloths, who were thumping huge tom-toms and chanting together. Being deaf, they were naturally unable to hear their own voices. The result was a weird, toneless melange of discordant sounds. Alma later wrote a paper for the *Journal of African Studies* in which she advanced the theory that the bone insertion practice "may" have had some influence on the character of their tribal music.

As we traced the eastern boundary of Bebopulu territory, Aguz called our attention to the sections that came in contact with the other three districts. As near as I could make out, the Bebopulu area, which I colored purple, extended south and then west around the southern end of the blue until it touched the green. I handed the tablet to Alma.

"You'll notice," I said, "that the blue is completely surrounded by other colors. There's no possible way a fifth area could border on it."

She showed the map to Aguz and they discussed it a moment. Then she gave it back to me. "He says he doesn't know what the districts look like from the sky, but he thinks you haven't got the shapes right." I glanced at Aguz. His face was immobile, but I had a strange feeling that inwardly he considered me something of an idiot.

The last area we entered was inhabited by natives I am at a total loss to describe. Their outstanding feature was, in fact, that they were completely nondescript. Anthropologists had studied the tribe for years and had been unable to isolate a single characteristic cultural trait.

Alma had been trying to work out a B.P.T. (Basic Personality Type) for the tribe, but so far her statistical analysis of the data had yielded nothing. There was no social organization; no kinship grouping; no standard rituals for birth, marriage, or death; no projective system of religious myths; and a total absence of folkways and mores. The tribe even lacked a name.

The natives, for the most part, ignored us as we walked through the village. They seemed to be moving about at random, doing nothing in particular. A few of the men carried spears, and occasionally a woman sauntered by balancing a water jar on her head.

I used yellow for my fifth color. As we passed the regions that touched on the green, red, and purple, I extended the yellow accordingly. When Aguz finally pointed off across a small creek and said that on the other side lived the Hiyikus (blue), I felt a strange, crawling sensation on my backbone.

"It can't be!" I shouted. "We must have walked through part of another district!" Alma translated to Aguz. He shook his head violently.

I was convinced, of course, that some mistake had been made. One of the territories must have two separate regions. Either that, or Aguz was misinformed about the boundaries. It *had* to be one or the other. When we got back to camp, shortly after sundown, Alma and I argued about it with great heat. She insisted I had lost, and demanded a check for the expenses of my trip.

I took off my sun helmet and blotted the bald portions of my head with a handkerchief. If there were only some way I could get an accurate map of the exact contours of the five regions! Of course they could be mapped by a careful ground survey, but then we had no surveying instruments, and I wouldn't know how to use them even if we had. Suddenly a wild idea entered my mind.

"Do you suppose," I said excitedly, "we could borrow some spray equipment in Monrovia?"

Alma squinted through the smoke from her cigarette and said she thought we could.

"If so," I went on, "we could spray spots of colored paint over each territory. Then if we took a color photograph from the air, we could see exactly how each district is shaped."

Alma broke into a big, charming smile. She thought the plan was wonderful. She had intended to map the island anyway, and this would be as quick a way to do it as any other. "We can even put the paint on my expense account," she said generously, waving her cigarette holder.

This brings my story up to the time at which I first began it. A building contractor in Monrovia let us borrow a dozen spray guns. I bought twenty thousand gallons of the cheapest paint I could get—a British product with a water base. Back on the island, we had no trouble in rounding up a crew of Hiyiku boys and teaching them how to use the spray equipment.

Aguz acted as foreman of the crew. We covered each territory, using the colors I had originally chosen for my map. Spraying every square foot of each region was too big a job, so we decided to spray spots of color, about twenty feet in diameter and at intervals of fifty yards. From the plane this would form a polka-dot pattern that would be easily recognizable. The job was simple in the open fields, but there was considerable difficulty in the jungle areas. The boys had to climb trees and spray paint on the topmost leaves. The natives enjoyed it immensely. They liked the brilliant colors.

I accompanied the crew each day to make sure every district was continuously spotted. There was no question that the first four areas we painted all had common borders because each border was clearly marked by the different colors on either side.

The fifth color was the crucial one!

We began spraying yellow on the twelfth day of our project. It touched the red, green, and purple all right, and as we began extending it toward the blue, I found myself in a state of high nervous tension. A few days previous I had developed an involuntary habit of jerking my head slightly. Now the tic had become so pronounced that I could control it only with great effort.

Our crew of native boys advanced slowly through the brush, the setting sun throwing long shadows ahead of them. A brightly colored parrot caught part of the spray and flapped away with shrill cries. A small brown snake, dripping with yellow paint, crawled out from a clump of vegetation. Suddenly I gripped Alma by the shoulder.

"Shades of Moebius!" I exclaimed hoarsely, my heart pounding. "I can see the blue spots from here!"

Alma's gray eyes danced with triumph. "Do you want to write that check out now?"

I sank down on the stump of a large tree and mopped away the perspiration that streamed down my face. My head jerked spasmodically. My temples throbbed. Above the monotonous hum of insects I could hear the frenzied, maddening rhythms of Bebopulu drums. Aguz stood by idly twirling his gold key and waiting for further orders.

I was at a loss to know what to think. Strictly speaking, it was not impossible that the five districts might actually have common borders. I knew there had been papers published proving four colors sufficient for districts up to 35 in number—but what if these "proofs" were later found to contain errors? If the island *did* refute the theorem, my discovery would be one of the great turning points in topology! I shook some white ants off my canteen and took a long drink of the cool water. I was beginning to feel better already.

A few days later, when the plane arrived on its weekly visit, we made arrangements for the camp photographer to take the aerial photographs. Unfortunately, the plane was a small one, with two open cockpits, so there was room only for the student with the camera. As soon as the pictures were made, the pilot was to bring the student back, then take me up to view the painted island.

I fanned myself nervously with my sun helmet and watched the plane circle slowly, then drop back to the clearing that was used as a landing field. The plane taxied to a stop and the photographer climbed out. I rushed over and prepared to take his place; but the pilot, a robust African who spoke good English, shook his head.

"Took longer than I expected," he said firmly. "I have to be back in Monrovia in half an hour. Sorry. Next week I'll take you up."

Nothing I could say would persuade him. When the plane took off a few minutes later, I turned to the photographer, my head twitching violently. "What did the island look like?"

He frowned. "I can't say exactly. The colors were funny shapes. I tried to sketch a map. Too complicated. Gave it up."

I asked if any of the colors were divided into separate parts, or if any area was surrounded by the others. He shook his head. "All solid colors. And all of them extend to the coast."

"Hmmmm," I said. "Interesting." Then the double-take hit me like a shotgun blast. I slapped my forehead and groaned.

Alma revived me by pouring cold water on my face from a gourd she'd carried from the camp's water hole. I sat up on the ground and held my head in my hands to keep it from jerking.

What had happened? I had realized suddenly that if each tribal area bordered on the sea, then the sea would border on all five districts. The ocean was a *sixth* color!

There were no facilities for developing color film either at the camp or in Monrovia. This meant the pictures had to wait until we got back to the States.

Three days later it began to rain. It rained steadily for the rest of the week. When the pilot returned to the island on his usual shuttle, he reported that the paint had completely washed away.

My interest in seeing the photographs had now reached such a feverish pitch that I could not wait until Alma completed her work on the island. I rode the plane to Monrovia, and returned to America on the first ship.

In New York I left the plates with a studio to be developed. When I picked them up, a few days later, my eyes were blood-streaked from lack of sleep.

"I'm afraid your photographer used the wrong type of filter," the clerk said, holding the transparencies up to the light for me. On each film the island was a solid mass of dark red! I took the pictures from him and stumbled into the street babbling incoherently.

My academic duties made it impossible to return to the island until the coming fall. I went back to Chicago and tried to tell my colleagues in the math department about the five districts. They smiled sadly and shook their heads. A professor at the University of West Virginia, they informed me, had just "proved" the theorem up to 83 districts. The dean suggested I take a month's rest. "You look haggard and run-down," he said.

By the end of the summer I had regained most of the weight I'd lost and my spirits began to improve. I checked the flight schedules to Monrovia. I wanted to visit that island and paint it again.

It was late September when I made the trip. Alma and the anthropology students had left the island several months earlier, and I had considerable difficulty finding the Hiyiku area. I finally made one of the Hiyikus understand that I wanted to see Aguz. He led me to a large hut on the outskirts of the village. Behind the hut was a tall, curious-shaped structure that glistened in the bright sunlight. It seemed to be made of large steel plates bolted together.

Aguz emerged from the hut followed by a short, chunky white man who—and then I felt my knees suddenly grow watery. No—it couldn't be! But it was! It was Dr. Stanislaw Slapenarski!

Aguz grinned and rushed over to support me, while the professor fanned my face with his helmet. He was looking better than I had ever seen him. The beard was just as red as I remembered it, and his face, including the bald head, was deeply tanned. He and Aguz helped me into the hut, where we all sat in comfortable chairs, and I downed a glass of brandy the professor handed me.

I haven't space to give in detail the amazing story Slapenarski had to tell. Let me summarize it by saying that in 1946, when the Polish topologist had made his sensational discovery of the no-sided surface, he had been extremely depressed by the publicity it received. He felt it was important he continue his unusual research in privacy, isolated from the curiosity of his colleagues, and from the prying eyes of reporters. In addition, Poland had gone behind the Iron Curtain and the administrators of the University of Warsaw had decided that the professor's views on topology exhibited "bourgeois idealistic tendencies."

"There was nothing to do but escape," Slapenarski said, in his familiar Polish accent. "I cabled false reports of my death to colleagues in England, France, and America. Then, with the aid of a forged passport, I managed to reach Monrovia."

After investigating several islands in the region, the professor had finally picked this one as ideal for a few years of uninterrupted research. He managed to acquire the Hiyiku dialect without much difficulty, and he had made Aguz, who had an excellent head for mathematics, his chief assistant. The island's five tribes at that time had no fixed areas in which to live, and were in a woeful state of deterioration and constant warfare. To maintain peace, it had been necessary to organize them into clearly defined districts.

"I had disproved the four-color theorem before I left Poland," he said, while I listened with intense inner excitement. "Dividing the island into five regions, all bordering each other, seemed a reasonable way to separate the tribes. With Aguz' help I marked out the boundaries, and the Hiyikus soon had the other natives under orderly control."

"Then you knew all about Dr. Bush and my previous visit here?" I asked.

"Of course. I'm dreadfully sorry, old man. But you see, I was in the midst of extremely important work, and I did not wish to be disturbed. Since Aguz was Dr. Bush's only contact on the island, it wasn't difficult to conceal my presence here. Naturally I could not afford to have you return to the States with the solution of the map problem. The island would have been deluged with photographers and newsreel men!"

"Then it was you," I said bitterly, "who ruined my film?"

"I'm afraid so, old chap. I had Aguz switch the filters on the camera. The rain, I must say, was providential. As soon as you left, of course, I altered the boundaries of the tribes."

"But—but how were they divided?" I said.

Slapenarski's little blue eyes twinkled. "Let me show you my laboratory," he said, rising.

A door at the back of the sitting room opened into a much larger room that was cluttered with huge sheets of colored cardboard, drafting equipment, books in cases along the walls, and large models of curious topological manifolds. I recognized a cross-cap,[6] a Tuckerman strip,[7] and a double Moebius band,[8] but the more complex models were unknown to me.

Then the professor led me out of the hut to a clearing in back. He waved his fat hand toward the gleaming steel structure I had observed earlier. "This has been my chief work for the past two years," he said. "It's a genuine Klein bottle."

[6] The cross-cap is a Moebius band deformed until the single edge becomes a circle, the band intersecting itself. An ordinary Moebius band is a one-sided, one-edged surface formed by giving a strip of paper one half-twist, then pasting the edges together.

[7] Dr. B. Tuckerman, after much folding and tucking, recently succeeded in producing a Moebius band with an edge in the form of a triangle, the strip remaining free of self-intersection. See *What Is Mathematics?* by Courant and Robbins, p. 262.

[8] The double Moebius is formed by putting two paper strips of equal length flat against each other, giving them a half-twist, and joining the paired ends. The result appears to be two Moebius bands, one nested within the other. In fact you can put a finger between the two bands and run it all the way around and back to the starting point, proving the bands separate. Nevertheless, the strip opens up into a single large band! Once the strip is opened, it is an exasperating puzzle to put it back again into the double Moebius form. Cutting the double strip down the middle produces two large rings interlocked. If the cut is begun a third of the way from the edge, cutting around the band twice to complete the cut, the result is even more surprising.

I shook my head in amazement.

There were two rope ladders leading to the top of the structure. We both climbed up and sat cautiously on the rounded rim. A current of cold air was rising from the opening.

"As you know," Slapenarski said, "the true Klein bottle is an aperture into the fourth dimension. It's like—how shall I put it?—like what a hole in a piece of paper would be to two-dimensional creatures on the paper's surface."

He explained this more fully. If you draw a two-dimensional bottle on a sheet of paper, he said, then imagine a portion of the bottle folded upward into the third dimension, you will see instantly that the content of the bottle is free to drop into our space. In similar fashion, he continued, while I tried to visualize it, the tubelike portion of a true Klein bottle twists in and out of the fourth dimension. The section that passes through higher space, although it remains closed in its three-dimensional cross-section, is actually open along the fourth coordinate. Anything dropped into the bottle is free at this point to fall in an infinite variety of directions into fourth-dimensional space.

I leaned forward warily and peered downward into the bottle's interior. The cold wind blew against my face. I could see nothing but a kind of grayish-green mist.

But my mind was still on the map problem. I questioned him again about it, and he seemed irritated at the change of subject. "The map problem is a trifle," he said, pulling on his red beard, "a mere trifle. Here—let me have your pad and pencil."

I took the tablet eagerly from my pocket and handed it to him. He drew some queer-looking geometrical shapes.

"If the map has no reducible configurations," the professor said, "such as non-triple vertices, multiple-connected regions, or rings formed by an even number of hexagons and pairs of adjacent pentagons, then—"

I am not sure I recall exactly what it was that happened. The horror has blurred my memory. Even now I find myself unable to write about it calmly.

A long black rod, the end curved into a hook like the tentacle of a gigantic insect, suddenly thrust itself upward from the dim interior of the bottle. The hook caught in the waist of Slapenarski's khaki shorts. He did not even have time to call out before he had been yanked downward into the bottle's misty depths.

I must have been in a state close to hysteria, but I do recall hearing Aguz shouting from the ground, and I remember looking down into the bottle's opening and seeing nothing but the swirling fog and feeling the icy upward rush of air. I yelled Slapenarski's name. But there were only the reverberations—like shouting into a deep well—then ominous silence. I had a wild fancy I could hear muttering voices—very faint, and in a strange tongue.

The rest can be briefly told. Word spread quickly among the Hiyikus about what had happened. That night a group of them invaded the premises, carried off the bottle, and tossed it over a nearby cliff. They believed that evil spirits were inside, and quite understandably wanted to destroy it.

Needless to say, no trace of the great topologist could be found in the smashed and twisted plates of steel. . . .

# BRUCE ELLIOTT

# The Last Magician

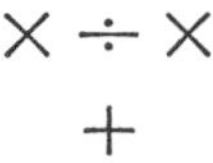

HE WAS the last one. I guess there's always something interesting about the last anything. The last dinosaur, the last auto, the last gas-powered plane, yes, he fits right into that museum of last things. He was the last magician.

He was good, too. I've seen the old celluloids of the great ones of the past, Houdini, Blackstone and Thurston, and he was like all of them rolled into one and more, much more. They functioned in a time when people still had a hankering to believe that there was such a thing as magic but he burst forth in our time like a nova. He revived interest in his hanky-panky art and he scared the hell out of people. He may have been a charlatan and paranoid and all the other things they called him, but he sure walloped the bejesus out of an audience, and that's something very few performers do these days.

I never knew why he chose the place he did for his debut except that it was good publicity-wise and that was something that he knew all about. He could sure pick his shots when it came to attracting public attention.

You know what vaudeville has become in our time—an intellectual's plaything, a cult for the avant garde. These vaudemanes sit and talk about tap dancers that were great hundreds of years ago and discuss crosstalk comics, whatever they were, and in general sit and drool about their dear, dear, dying art form.

I don't know much about art forms and I have a sneaking feeling that anything that can't support itself by public interest doesn't amount to much. Certainly vaudeville would be nonexistent if it wasn't subsidized by these cultists. But I made my living cooking up props for these phony shows and that was good enough for me. Until

I joined Duneen as a prop man I had always worked with my hands and you know that means I'm good because you have to be better than a machine to get a license to work with your hands today.

But I was telling you about Duneen. He walked out on the little stage where we put on our "vaudevilles," completely unannounced. His appearance sure made everyone sit up and take notice. Heaven knows where he got his outfit because it was a real costume piece. Black cape swirling around his tall, lean frame; a curious kind of butterfly-shaped thing at his neck that was surrounded with a high white band, a completely nonfunctional jacket that was cut away in the front and dropped down like tails in the rear and a shirt that looked as if it was made of some stiff plastic. It would have looked funny on anyone else but it didn't on him.

I suppose the hair he had on his lip and chin was fake because all males have their face hair extirpated at puberty now, but I never saw him without it. He called the hair a mustache and goatee and it did strange things to his hollow-cheeked face.

He walked out to the center of the stage and bowed obsequiously to the handful of avant gardists that made up the audience. But somehow, even the bow, even the mock humility was an insult. It was as though he was just pretending to be humble because he knew he was superior. He could get under your skin like that in a million ways but I didn't learn that till later.

I could hear a little rustle in the audience as they looked through their programs trying to figure out who Duneen was. They didn't have much time for that though, because as he bowed he swept off his cape and gracefully showed both sides of it.

His curious lip-tilted grimace that was halfway between a smile and a sneer appeared as he draped the cape over his arm. Suddenly there was a form under it. When he whipped the cloth away a Martian girl, naked, stood shyly revealed. Duneen looked at the audience out of the corners of his eyes as though trying to gauge the effect and then plucked a wand out of the air. This is a long black stick with white tips. It is an adjunct that old-time magicians always used.

Gesturing at the girl with the "wand," he then snapped his fingers. Suddenly a brassiere appeared clothing her breasts. Another snap of those long, thin fingers and her legs and thighs were covered. Then he gestured around her with the "wand" and she was fully dressed. The cloth seemed to come from nowhere, seemed to be produced at the tip of the "wand."

From then on she assisted him as you have seen her do on TV. The only reason I'm telling you about this first opening is that they never let him repeat it on the air. The Martian ambassador complained and there was some kind of a stink; I don't know what it was all about, but Duneen never started his show that way again.

You remember the rest of his act, of course; the sawing of the Martian girl in half with a G ray and her restoration after you would have sworn she was dead. The way he would cause her to vanish from an hermetically sealed rocket blast tube and the way he produced her from a previously shown empty Liane lizard shell. All these things became household words and that was just the trouble.

Just because he was the last of the magicians, just because he had such a terrific effect on show business, he had to keep topping himself. He had to keep inventing newer and more amazing tricks and it almost drove him crazy.

Then there was the other reason which had been true since the beginning of TV. TV is a bottomless maw into which entertainment is shoveled only to vanish like one of Duneen's tricks. Centuries ago, when the TV audience was just made up of millions of people, you could repeat yourself once in a while, I suppose, and figure that not everyone had caught you the first time. But now, when the audience is up in the hundreds of millions, the problem has become so bad that lots of performers crack under the strain.

The old-time magicians used to meet their audiences bit by bit through the years, and if there was any overlapping it didn't matter very much. But now, today, you meet all the people in the world with one performance.

I've read in the old magic textbooks that magicians could, and did, do the same tricks over and over for the length of their professional careers. Imagine that!

But Duneen, of course, could never repeat himself, even once. He had to keep inventing more and more exciting variations on his basic tricks.

That was where I came in, me and my capable hands. I guess maybe I wouldn't have helped him if it hadn't been for Aydah, his Martian girl assistant—but I felt sorry for her. He was nasty to her most of the time, but he was at his worst when he was racking his brain trying to cook up a new pseudo-miracle.

I heard her crying one day. Heard it right through even the thick walls of the dressing room at the TV studio. You could say it was none of my business, but I busted in anyway and said, "Can I help, Aydah?"

You wouldn't think a girl seven feet tall and so thin that her veins stood out like cords could look wistful and appealing, but she did. Her bright red eyes were glistening with tears which she certainly could ill afford to waste, considering how dehydrated Martians are.

She said, "What can you do? What can anyone do?" Luckily she was sitting down, sort of scrunched over because I put her head on my shoulder and patted the long, thin white hair, which I certainly would not have been able to do without a ladder if she'd been standing, and said, "Tell me about it."

"Mr. Barrow," she gulped, "I guess I sort of love him or I wouldn't stay on—but how can I love and hate someone at the same time?"

I patted her head and was silently sorry for her.

She asked, "Don't you know? I've read all the Earth books I could find, all that have anything to do with love and I can't find any answer." She sobbed, "They don't explain it at all. Can't you tell me?"

That was a poser all right. I'm past the age where sex or love or any of that sort of nonsense means very much to me, but I have a good memory. . . .

"Whatever possessed you to fall for an Earthman, Aydah?" It was a stupid question but I was just making conversation.

She lowered her head and rested it on my chest. I kept patting it sort of ineffectually while she talked. "I don't really know. He came along when I was the right age and Mother had always kept me away from Martian boys. She kept saying I wasn't old enough. She didn't see any danger in an Earthman, I guess. But Duneen isn't fat like you, Mr. Barrow, or like most Earth people. He's almost as thin and handsome as a Martian. And he can talk so beautifully—when he wants to." She was off in racking sobs again.

That was when Duneen stalked in. He was in high, low and medium dudgeon. He said, "Why you—Martian guttersnipe! I take you in and this is the way you behave the first time my back is turned. Carrying on with an old man! Why you . . ."

He looked all set to beat her up so I intervened. I said, "Look,

Duneen, you know that I've come up with some good ideas for your show."

He nodded. At least I had his attention. I went on quickly, "I think I have a brand-new idea for an escape."

Jealousy faded before his interest in a new trick. He asked, "What's the gag?"

"You've escaped from every kind of gadget that anyone could think up. You've challenged people to think of restraints that will hold you for more than five minutes, right?"

"Of course," Duneen said impatiently. "I've escaped from things that would have killed that old-timer, Houdini!" He grunted. "That old faker! I get mad every time I read about him!"

He did too. He seemed to be furious because he had come too late in time to match wits with the great magicians of the earlier days. He felt, and I guess he was right, that he could have topped any of them.

"What is it?" he asked impatiently, turning back to face Aydah.

I said quickly, "How about escaping from a Klein bottle?"

"What? What's that?"

I sighed. Sometimes his stupidity about anything outside of his own field appalled me. I made it as simple as I could. "Look," I said, picking up a narrow strip of paper, "you know what a Moebius strip is?"

He looked unsure so I glued one end of the strip to the other end making the half twist in the paper that has to be made in order for the topological principle to work. Using a pencil I showed him how a line could be drawn on both sides of the paper despite the fact that I didn't lift the pencil from the paper. I said, "See? It's a one-sided figure!"

He grunted. "Oh that!" He picked up a pair of scissors and cut around the loop of paper. It formed into two interlocked circles, of course. He said, "This is the Afghan bands. Why didn't you say so?"

"Maybe that's what magicians used to call it," I said, "but it's a Moebius strip and it will help . . ."

He was scowling now, all thought of Aydah gone from his mind. He asked, "What's all this got to do with me? I can't escape from a strip of paper. That's ridiculous!"

"No, of course not. But if you think of this strip of paper as a two-dimensional object that has strange properties because of the twist in it, which is in the third dimension, it will help you to think about the Klein bottle."

He raised his eyebrows.

"Look," I said, "a Klein bottle is a fourth-dimensional equivalent of the Moebius strip. Picture a bottle made out of a hard rubbery substance. Now bend the neck of the bottle down and around, and push the mouth of the bottle through the side of the bottle without breaking the surface of the bottle."

He really wasn't too stupid. He said, "That's the point at which it goes through the fourth dimension, eh?"

"Yes, now suppose I made up a bottle like that big enough for you to get into . . ."

"So what's so good about escaping from a bottle? That has no drama, no excitement!"

"You don't get it! According to topological laws which were proved the first time they made a real Klein bottle fifty years ago, a fly walking on the surface of the bottle is on the inside-outside of the bottle and can never get in or out of the bottle. Any schoolboy knows that!"

He whistled through his teeth. "I think you have something there. Not that the basic idea is much good, but I'll build on it. I'll make this the most sensational escape that has ever been done. Houdini! Phooey!" A sudden thought stopped him. "What's the gaff?"

I said, "Huh?" But I knew what he meant. He always irritated me, using show business terms that had been obsolete for many years, although I've noticed lately that he has me doing it too.

"What's the gaff?" he repeated. "How do I escape the fate of the fly?"

"You're not thinking, Duneen. If you ever climbed into a real Klein bottle that would be the end of you. You'd be alive-dead. Halfway between here and the fourth-dimensional world—you'd be stranded!"

"So?" he asked.

"So we have to rig up a substitute bottle. A fake."

"Okay, it's a deal. You get to work on it." He turned his attention back to Aydah. He said, "Now you, listen to me!" She cowered away from him.

She had to listen to him. I didn't. I left but I was mad. Bullying Aydah was about on a par with kicking a sick puppy. If I could, I would have taken a punch at him; not that it would have done any good.

He could sneer all he wanted to at Houdini and the other old-timers but he had learned their lesson well. His publicity on the Klein-bottle escape was a masterpiece. By the time I had constructed the two bottles, the real and the fake one, he had everyone talking about Klein bottles and how foolhardy he was, how he was defying the most dreadful fate a man had ever faced. He planted pieces in the news about topology. He had planes drop hundreds of thousands of Moebius strips and each strip had DUNEEN DEFIES DEATH! lettered on it along with instructions about the strip. He bombarded the press services with handouts. He challenged Miklav and Ronner, the two top topologists of the day, to figure out how he would escape. He bet them 10,000 credits that he would escape in five minutes, with a proviso that he would pay a thousand credits a minute to their favorite charity for every minute over five that he was stuck in the bottle.

The harder he worked the worse he treated Aydah. I had to keep out of the way or I would have hung a punch on his long, aquiline nose.

It seemed as if every time I turned around I'd find her hiding in some corner, crying. The loss of water through her tears began to tell on her. I finally had to call in a doctor and have some saline solution injected intravenously or she would have just faded away. It was when she was stretched out getting the intravenous that I first noticed that her ordinarily concave stomach was getting a little convex.

I guess that was when I began to get really mad at Duneen. Mad enough to do something about the whole bloody mess . . .

But she never really complained, not out loud anyway. That one outburst to me was all. She would just mope around and look at Duneen hopefully, and then her eyes would fill up with tears and off she'd go for another quiet cry.

I tell you it got me down. But there was nothing I could do, not even when I found out what was back of it all. I spotted Duneen one night with another girl, an Earth girl, but I couldn't see where it would do any good to tell Aydah that. So I just kept busy on my props, getting everything ready and keeping my fingers crossed.

If you were anywhere within eyesight of a TV set that night you saw what happened, at least from out front. But I know what happened backstage, and that's what I wanted to tell you about.

Everything went off like clockwork and you can believe me when I say that he was magnificent. With all his faults, with all his pettiness, despite his charlatanry, or maybe because of it, he was great. The last of the magicians and the greatest!

Naturally, he didn't open with the escape. That was to be his climax. He prefaced it with little run-of-the-mill items like an endless production of Martian geezers, those cute little six-legged creatures with the red eyes and white hair. They always reminded me of Aydah, and that night I was more aware of the resemblance as he kept reaching into his tall hat and producing the little things as though the supply was endless. Then it was pure poetry when he plucked obsolete coins of every denomination out of the air and sent them clattering into a metal bowl. You know: parlor tricks, simple little things, but he did them with such an air!

Backstage the technicians kept a wary eye on the real Klein bottle which I had ready. I could see that they wanted no part of it or of the fate of the man who was supposed to escape from it.

When Duneen was sure that he had milked every bit of suspense out of his act he stopped and held up his hands in that corny, theatrical gesture of his and said, "Ladies and gentlemen, next—I present the challenge escape of all time! I shall enter a Klein bottle. . . ."

He gestured at it as it was wheeled on stage. There was no sound as the stagehands placed a three-fold screen around the bottle. Duneen went on, "I will escape from that bottle in five minutes or . . ." He was a good enough showman not to finish the sentence.

He had Miklav and Donner come on stage and examine the bottle. They seemed oddly out of place, these men of science, these topologists, as they examined the bottle.

Duneen said, "Gentlemen, do you agree that the bottle is a true Klein bottle?" They nodded.

Duneen went off stage. He was sure enough of himself to leave the stage empty while he changed into trunks. His excuse was to show that he had no gadgets on his person to aid him. He always performed his escapes that way. But I never thought that this was the real reason he stripped. I think he liked to hear the shocked gasp when people saw his skeletally thin frame. Of all Earthmen I've ever seen, he came closest to looking like a Martian. Seeing him that way I could understand a little better why Aydah had fallen in love with him.

I was off stage, left. I had nothing to do but keep an eye on things. Nothing much could go wrong because I had decided that the best way to switch the real and the fake Klein bottles would, after all, be the simplest way. I had made two trap doors in the stage. I don't suppose anyone has used traps for tricks for centuries. That's why I was sure the hoary old gag would fool the audience. Duneen agreed with me and he was never wrong about what would fool people.

The arrangement was merely this. The real Klein bottle was on stage and would stay there until the experts had examined it and pronounced it to be indubitably what it was, a fourth-dimensional bottle. Once they had pronounced it legitimate, Duneen would conceal it behind a three-fold screen; pressure on a button would activate the trap doors. The real bottle would sink out of sight. A fake Klein bottle, which looked real enough but did not have the properties of the topological figure, would rise up to replace the genuine one.

As you can see, the mechanics of the trick were a cinch. But, according to Duneen, that was the real secret of good magic. Complexity, he maintained, is no good. People can dope it out. You must use a simple device, so simple that your audience discards it as a possibility just because of its simplicity.

Duneen stood next to me in the wings, breathing deeply, bracing himself for his appearance on stage. The button that made the trap doors work was near us, on the wall. He pressed the button. Aydah ran over to join us as, outside, on stage, the announcer was saying, "And now—we have the honor and privilege of presenting . . ." There was a long drum roll and then, "Duneen!"

That was the cue. He stalked out on stage. Aydah was next to me. We both watched him. Duneen was bowing to the audience. He blew a kiss to a girl who was sitting down front. She was the Earth girl I had seen him with. I was near enough to Aydah to feel her thin body stiffen. Then she did know about Duneen and . . .

On stage center Duneen motioned for the stagehands to remove the screen that had masked the man-sized bottle. He gestured at it. His grin was at its most sardonic as he lifted one of his spidery legs and placed it around the shoulder of the bottle. The stagehands stood ready with the screen and as he nodded to them they stepped forward with it. He lifted his other leg preparatory to mounting the bottle like a horse.

Aydah shivered and then sobbed, "I can't . . . I can't let him do it!" The screen was almost around the magician now. She reached over my shoulder and tried to press the button that would switch the bottle.

"What are you doing?" I asked.

"I . . ." Her eyes were frantic. "I can't do it! I switched the bottles before! That's the *real* Klein bottle he's . . ."

It was too late to press the button. She said hastily, "I'll go tell him to stall! Then when the screen hides him completely, you press the button and switch back the fake bottle! How could I have been so cruel!" She ran out on stage.

Darting to his side she whispered to him. Even then, with the eyes of the world on him, he almost cuffed her. I saw his hand start up, saw her back away, before he caught himself and remembered where he was. He managed to turn his grimace of hatred into a smile as he turned to face the audience.

"Ladies and gentlemen, my 'invaluable' assistant tells me that there are some reporters backstage who would like to be out here as a committee. I extend my welcome to them!"

It was a good stall. I don't think anyone knew what had really happened. The stagehands surrounded the bottle with the screen while Duneen bowed to the reporters.

Aydah ran to my side. "Press the button."

She watched me while I did so and then turned and made a motion for Duneen to proceed. The side of his face to the audience was smiling, but there was black and bitter loathing in his eyes when he turned away.

He faced the bottle again. The screen was brought forward. Mounting the bottle, his arms and legs straddled the shoulder of it. Then, as he allowed himself to slide down toward the spot where the mouth of the bottle went through the side of it, a curious thing happened.

He seemed to become rubbery. One moment he was all on the outside of the bottle, the next, a cross-section of him seemed to be inside it. That was all anyone saw as the screen cut off the view.

Aydah sobbed at my side. "Let him go to her. I have no hold on him. We're not married . . . we could never marry, not with the law about miscegenation between Mars and Earth people. Let her have him."

"None," I agreed, "except for the fact that he has condemned you to death!"

Involuntarily she looked down at her little belly. Then she looked at me. "You knew?"

"Sure, I could see you were pregnant a month ago. And there's no escape from the death penalty for miscegenation." I patted her shoulder. "He should have had you aborted while there was still time."

"It's too late now," she said and turned her back. I knew that as well as she did.

On stage the reporters were eying their watches. The music, keyed for suspense, was getting nerve-racking as the minutes dragged by. The audience became restive. The two professors of topology looked frightened. One of them, I think it was Miklav, broke away from a friend who tried to restrain him. Miklav shouted, "What do I care about any bet! That man is in trouble!"

He shoved aside the screen and, of course, he was right. Duneen was in real bad trouble. He was half in and half out of the Klein bottle. He was on the inside-outside, never-come-right-side of the bottle. There he was, and there he is now. In the museum with all the other last things. And there he'll stay. They can't break the bottle because that would divide him. And since they can't break the bottle there he will remain, not alive and not dead—suspended midway between here and there. Wherever *there* is in the fourth dimension.

It isn't very pretty. But then neither was what he did to Aydah. I might have felt just a little pity for him, but I saw her die. She killed herself just before the authorities got around to it.

I knew she would have to die. That was why I had pressed the button that switched the bottles the first time, before she ever did. That canceled out the later switch when she thought she was saving him. . . . It made an odd sequence.

Get it? The real bottle was up there on the stage when the topologists looked at it. I switched it for the fake one so, when Duneen made *his* switch, it was the real one that came up! Aydah almost screwed up the works when she pulled her switch and brought the fake bottle back up on stage. It turned out okay, though. She thought the real bottle was up there and when she begged me to make the change—the real Klein bottle was ready and waiting for Duneen!

Sometimes when I go to the museum of last things to look at him, I think of the old stories about evil genies and the way they were stuffed into bottles. I guess I must be getting old; lately I've taken to wondering about King Solomon. He knew so much, I wonder if he knew about Klein bottles. . . .

A. J. DEUTSCH

# A Subway Named Moebius

IN A COMPLEX and ingenious pattern, the subway had spread out from a focus at Park Street. A shunt connected the Lechmere line with the Ashmont for trains southbound, and with the Forest Hills line for those northbound. Harvard and Brookline had been linked with a tunnel that passed through Kenmore Under, and during rush hours every other train was switched through the Kenmore Branch back to Egleston. The Kenmore Branch joined the Maverick Tunnel near Fields Corner. It climbed a hundred feet in two blocks to connect Copley Over with Scollay Square; then it dipped down again to join the Cambridge line at Boylston. The Boylston shuttle had finally tied together the seven principal lines on four different levels. It went into service, you remember, on March 3. After that, a train could travel from any one station to any other station in the whole system.

There were two hundred twenty-seven trains running the subways every weekday, and they carried about a million and a half passengers. The Cambridge-Dorchester train that disappeared on March 4 was Number 86. Nobody missed it at first. During the evening rush, the traffic was a little heavier than usual on that line. But a crowd is a crowd. The ad posters at the Forest Hills yards looked for 86 about 7:30, but neither of them mentioned its absence until three days later. The controller at the Milk Street Cross-Over called the Harvard checker for an extra train after the hockey game that night, and the Harvard checker relayed the call to the yards. The dispatcher there sent out 87, which had been put to bed at ten o'clock, as usual. He didn't notice that 86 was missing.

It was near the peak of the rush the next morning that Jack O'Brien, at the Park Street Control, called Warren Sweeney at the Forest Hills yards and told him to put another train on the Cambridge run. Sweeney was short, so he went to the board and scanned it for a spare train and crew. Then, for the first time, he noticed that Gallagher had not checked out the night before. He put the tag up and left a note. Gallagher was due on at ten. At ten-thirty, Sweeney was down looking at the board again, and he noticed Gallagher's tag still up, and the note where he had left it. He groused to the checker and asked if Gallagher had come in late. The checker said he hadn't seen Gallagher at all that morning. Then Sweeney wanted to know who was running 86? A few minutes later he found that Dorkin's card was still up, although it was Dorkin's day off. It was 11:30 before he finally realized that he had lost a train.

Sweeney spent the next hour and a half on the phone, and he quizzed every dispatcher, controller, and checker on the whole system. When he finished his lunch at 1:30, he covered the whole net again. At 4:40, just before he left for the day, he reported the matter, with some indignation, to Central Traffic. The phones buzzed through the tunnels and shops until nearly midnight before the general manager was finally notified at his home.

It was the engineer on the main switchbank who, late in the morning of the sixth, first associated the missing train with the newspaper stories about the sudden rash of missing persons. He tipped off the *Transcript*, and by the end of the lunch hour three papers had extras on the streets. That was the way the story got out.

Kelvin Whyte, the general manager, spent a good part of that afternoon with the police. They checked Gallagher's wife, and Dorkin's. The motorman and the conductor had not been home since the morning of the fourth. By mid-afternoon, it was clear to the police that three hundred and fifty Bostonians, more or less, had been lost with the train. The System buzzed, and Whyte nearly expired with simple exasperation. But the train was not found.

Roger Tupelo, the Harvard mathematician, stepped into the picture the evening of the sixth. He reached Whyte by phone, late, at his home, and told him he had some ideas about the missing train. Then he taxied to Whyte's home in Newton and had the first of many talks with him about Number 86.

Whyte was an intelligent man, a good organizer, and not without imagination. "But I don't know what you're talking about!" he expostulated.

Tupelo was resolved to be patient. "This is a very hard thing for *anybody* to understand, Mr. Whyte," he said. "I can see why you are puzzled. But it's the only explanation. The train has vanished, and the people on it. But the System is closed. Trains are conserved. It's somewhere on the System!"

Whyte's voice grew louder again. "And I tell you, Dr. Tupelo, that train is *not* on the System! It is *not!* You can't overlook a seven-car train carrying four hundred passengers. The System has been combed. Do you think I'm trying to *hide* the train?"

"Of course not. Now look, let's be reasonable. We know the train was en route to Cambridge at 8:40 A.M. on the fourth. At least twenty of the missing people probably boarded the train a few minutes earlier at Washington, and forty more at Park Street Under. A few got off at both stations. And that's the last. The ones who were going to Kendall, to Central, to Harvard—they never got there. The train did not get to Cambridge."

"I know that, Dr. Tupelo," Whyte said savagely. "In the tunnel under the river, the train turned into a boat. It left the tunnel and sailed for Africa."

"No, Mr. Whyte. I'm trying to tell you. It hit a node."

Whyte was livid. "What is a node!" he exploded. "The System keeps the tracks clear. Nothing on the tracks but trains, no nodes left lying around—"

"You still don't undcrstand. A nodc is not an obstruction. It's a singularity. A pole of high order."

Tupelo's explanations that night did not greatly clarify the situation for Kelvin Whyte. But at two in the morning, the general manager conceded to Tupelo the privilege of examining the master maps of the System. He put in a call first to the police, who could not assist him with his first attempt to master topology, and then, finally, to Central Traffic. Tupelo taxied down there alone and pored over the maps till morning. He had coffee and a snail, and then went to Whyte's office.

He found the general manager on the telephone. There was a conversation having to do with another, more elaborate, inspection of the Dorchester-Cambridge tunnel under the Charles River. When the conversation ended, Whyte slammed the telephone into its cradle and glared at Tupelo. The mathematician spoke first.

"I think probably it's the new shuttle that did this," he said.

Whyte gripped the edge of his desk and prowled silently through his vocabulary until he had located some civil words. "Dr. Tupelo," he said, "I have been awake all night going over your theory. I don't understand it at all. I don't know what the Boylston shuttle has to do with this."

"Remember what I was saying last night about the connective properties of networks?" Tupelo asked quietly. "Remember the Moebius band we made—the surface with one face and one edge? Remember this—?" and he removed a little glass Klein bottle from his pocket and placed it on the desk.

Whyte sat back in his chair and stared wordlessly at the mathematician. Three emotions marched across his face in quick succession—anger, bewilderment, and utter dejection. Tupelo went on.

"Mr. Whyte, the System is a network of amazing topological complexity. It was already complex before the Boylston shuttle was installed, and of a high order of connectivity. But this shuttle makes the network absolutely unique. I don't fully understand it, but the situation seems to be somehing like this: the shuttle has made the connectivity of the whole System of an order so high that I don't know how to calculate it. I suspect the connectivity has become infinite."

The general manager listened as though in a daze. He kept his eyes glued to the little Klein bottle.

"The Moebius band," Tupelo said, "has unusual properties because it has a singularity. The Klein bottle, with two singularities, manages to be inside of itself. The topologists know surfaces with as many as a thousand singularities, and they have properties that make the Moebius band and the Klein bottle both look simple. But a network with infinite connectivity must have an infinite number of singularities. Can you imagine what the properties of that network could be?"

After a long pause, Tupelo added: "I can't either. To tell the truth, the structure of the System, with the Boylston shuttle, is completely beyond me. I can only guess."

Whyte swiveled his eyes up from the desk at a moment when anger was the dominant feeling within him. "And you call yourself a mathematician, Professor Tupelo!" he said.

Tupelo almost laughed aloud. The incongruousness, the absolute foolishness of the situation, all but overwhelmed him. He smiled thinly and said: "I'm no topologist. Really, Mr. Whyte, I'm a tyro in the field—not much better acquainted with it than you are. Mathematics is a big pasture. I happen to be an algebraist."

His candor softened Whyte a little. "Well, then," he ventured, "if you don't understand it, maybe we should call in a topologist. Are there any in Boston?"

"Yes and no," Tupelo answered. "The best in the world is at Tech."

Whyte reached for the telephone. "What's his name?" he asked. "I'll call him."

"Merritt Turnbull. He can't be reached. I've tried for three days."

"Is he out of town?" Whyte asked. "We'll send for him—emergency."

"I don't know. Professor Turnbull is a bachelor. He lives alone at the Brattle Club. He has not been seen since the morning of the fourth."

Whyte was uncommonly perceptive. "Was he on the train?" he asked tensely.

"I don't know," the mathematician replied. "What do you think?"

There was a long silence. Whyte looked alternately at Tupelo and at the glass object on the desk. "I don't understand it," he said finally. "We've looked everywhere on the System. There was no way for the train to get out."

"The train didn't get out. It's still on the System," Tupelo said.

"Where?"

Tupelo shrugged. "The train has no real 'where.' The whole System is without real 'whereness.' It's double-valued, or worse."

"How can we find it?"

"I don't think we can," Tupelo said.

There was another long silence. Whyte broke it with a loud exclamation. He rose suddenly, and sent the Klein bottle flying across the room. "You are crazy, Professor!" he shouted. "Between midnight tonight and 6:00 A.M. tomorrow, we'll get every train out of the tunnels. I'll send in three hundred men, to comb every inch of the tracks—every inch of the one hundred eighty-three miles. We'll find the train! Now, please excuse me." He glared at Tupelo.

Tupelo left the office. He felt tired, completely exhausted. Mechanically he walked along Washington Street toward the Essex Station. Halfway down the stairs he stopped abruptly, looked around him slowly. Then he ascended again to the street and hailed a taxi. At home, he helped himself to a double shot. He fell into bed.

At 3:30 that afternoon he met his class in "Algebra of Fields and Rings." After a quick supper at the Crimson Spa, he went to his apartment and spent the evening in a second attempt to analyze the connective properties of the System. The attempt was vain, but the mathematician came to a few important conclusions. At eleven o'clock he telephoned Whyte at Central Traffic.

"I think you might want to consult me during tonight's search," he said. "May I come down?"

The general manager was none too gracious about Tupelo's offer of help. He indicated that the System would solve this little problem without any help from harebrained professors who thought that whole subway trains could jump off into the fourth dimension. Tupelo submitted to Whyte's unkindness, then went to bed. At about 4:00 A.M. the telephone awakened him. His caller was a contrite Kelvin Whyte.

"Perhaps I was a bit hasty last night, Professor," he stammered. "You may be able to help us after all. Could you come down to the Milk Street Cross-Over?"

Tupelo agreed readily. He felt none of the satisfaction he had anticipated. He called a taxi, and in less than half an hour was at the prescribed station. At the foot of the stairs, on the upper level, he saw that the tunnel was brightly lighted, as during normal operation of the System. But the platforms were deserted except for a tight little knot of seven men near the far end. As he walked toward the group, he noticed that two were policemen. He observed a one-car train on the track beside the platform. The forward door was open, the car brightly lit and empty. Whyte heard his footsteps and greeted him sheepishly.

"Thanks for coming down, Professor," he said, extending his hand. "Gentlemen, Dr. Roger Tupelo, of Harvard. Dr. Tupelo, Mr. Kennedy, our chief engineer; Mr. Wilson, representing the mayor; Dr. Gannot, of Mercy Hospital." Whyte did not bother to introduce the motorman and the two policemen.

"How do you do," said Tupelo. "Any results, Mr. Whyte?"

The general manager exchanged embarrassed glances with his companions. "Well . . . yes, Dr. Tupelo," he finally answered. "I think we do have some results, of a kind."

"Has the train been seen?"

"Yes," said Whyte. "That is, practically seen. At least, we know it's somewhere in the tunnels." The six others nodded their agreement.

Tupelo was not surprised to learn that the train was still on the System. After all, the System was closed. "Would you mind telling me just what happened?" Tupelo insisted.

"I hit a red signal," the motorman volunteered. "Just outside the Copley junction."

"The tracks have been completely cleared of all trains," Whyte explained, "except for this one. We've been riding it, all over the System, for four hours now. When Edmunds, here, hit a red light at the Copley junction, he stopped, of course. I thought the light must be defective, and told him to go ahead. But then we heard another train pass the junction."

"Did you see it?" Tupelo asked.

"We couldn't see it. The light is placed just behind a curve. But we all heard it. There's no doubt the train went through the junction. And it must be Number 86, because our car was the only other one on the tracks."

"What happened then?"

"Well, then the light changed to yellow, and Edmunds went ahead."

"Did he follow the other train?"

"No. We couldn't be sure which way it was going. We must have guessed wrong."

"How long ago did this happen?"

"At 1:38, the first time—"

"Oh," said Tupelo, "then it happened again later?"

"Yes. But not at the same spot, of course. We hit another red signal near South Station at 2:15. And then at 3:28—"

Tupelo interrupted the general manager. "Did you see the train at 2:15?"

"We didn't even hear it that time. Edmunds tried to catch it, but it must have turned off onto the Boylston shuttle."

"What happened at 3:28?"

"Another red light. Near Park Street. We heard it up ahead of us."

"But you didn't see it?"

"No. There is a little slope beyond the light. But we all heard it. The only thing I don't understand, Dr. Tupelo, is how that train could run the tracks for nearly five days without anybody seeing—"

Whyte's word trailed off into silence, and his right hand went up in a peremptory gesture for quiet. In the distance, the low metallic thunder of a fast-rolling train swelled up suddenly into a sharp, shrill roar of wheels below. The platform vibrated perceptibly as the train passed.

"Now we've got it!" Whyte exclaimed. "Right past the men on the platform below!" He broke into a run toward the stairs to the lower level. All the others followed him, except Tupelo. He thought he knew what was going to happen. It did. Before Whyte reached the stairs, a policeman bounded up to the top.

"Did you see it, now?" he shouted.

Whyte stopped in his tracks, and the others with him.

"Did you see that train?" the policeman from the lower level asked again, as two more men came running up the stairs.

"What happened?" Wilson wanted to know.

"Didn't you see it?" snapped Kennedy.

"Sure not," the policeman replied. "It passed through up here."

"It did *not*," roared Whyte. "Down there!"

The six men with Whyte glowered at the three from the lower level. Tupelo walked to Whyte's elbow. "The train can't be seen, Mr. Whyte," he said quietly.

Whyte looked down at him in utter disbelief. "You heard it yourself. It passed right below—"

"Can we go to the car, Mr. Whyte?" Tupelo asked. "I think we ought to talk a little."

Whyte nodded dumbly, then turned to the policeman and the others who had been watching at the lower level. "You really didn't see it?" he begged them.

"We heard it," the policeman answered. "It passed up here, going that way, I think," and he gestured with his thumb.

"Get back downstairs, Maloney," one of the policemen with Whyte commanded. Maloney scratched his head, turned, and disappeared below. The two other men followed him. Tupelo led the original group to the car beside the station platform. They went in and took seats, silently. Then they all watched the mathematician and waited.

"You didn't call me down here tonight just to tell me you'd found the missing train," Tupelo began, looking at Whyte. "Has this sort of thing happened before?"

Whyte squirmed in his seat and exchanged glances with the chief engineer. "Not exactly like this," he said, evasively, "but there have been some funny things."

"Like what?" Tupelo snapped.

"Well, like the red lights. The watchers near Kendall found a red light at the same time we hit the one near South Station."

"Go on."

"Mr. Sweeney called me from Forest Hills at Park Street Under. He heard the train there just two minutes after we heard it at the Copley junction. Twenty-eight track miles away."

"As a matter of fact, Dr. Tupelo," Wilson broke in, "several dozen men have seen lights go red, or have heard the train, or both, inside of the last four hours. The thing acts as though it can be in several places at once."

"It can," Tupelo said.

"We keep getting reports of watchers seeing the thing," the engineer added. "Well, not exactly seeing it, either, but everything except that. Sometimes at two or even three places, far apart, at the same time. It's sure to be on the tracks. Maybe the cars are uncoupled."

"Are you really sure it's on the tracks, Mr. Kennedy?" Tupelo asked.

"Positive," the engineer said. "The dynamometers at the power house show that it's drawing power. It's been drawing power all night. So at 3:30 we broke the circuits. Cut the power."

"What happened?"

"Nothing," Whyte answered. "Nothing at all. The power was off for twenty minutes. During that time, not one of the two hundred fifty men in the tunnels saw a red light or heard a train. But the power wasn't on for five minutes before we had two reports again—one from Arlington, the other from Egleston."

There was a long silence after Whyte finished speaking. In the tunnel below, one man could be heard calling something to another. Tupelo looked at his watch. The time was 5:20.

"In short, Dr. Tupelo," the general manager finally said, "we are compelled to admit that there may be something in your theory." The others nodded agreement.

"Thank you, gentlemen," Tupelo said.

The physician cleared his throat. "Now about the passengers," he began. "Have you any idea what—?"

"None," Tupelo interrupted.

"What should we do, Dr. Tupelo?" the mayor's representative asked.

"I don't know. What can you do?"

"As I understand it from Mr. Whyte," Wilson continued, "the

train has . . . well, it has jumped into another dimension. It isn't really on the System at all. It's just gone. Is that right?"

"In a manner of speaking."

"And this . . . er . . . peculiar behavior has resulted from certain mathematical properties associated with the new Boylston shuttle?"

"Correct."

"And there is nothing we can do to bring the train back to . . . uh . . . this dimension?"

"I know of nothing."

Wilson took the bit in his teeth. "In this case, gentlemen," he said, "our course is clear. First, we must close off the new shuttle, so this fantastic thing can never happen again. Then, since the missing train is really gone, in spite of all these red lights and noises, we can resume normal operation of the System. At least there will be no danger of collision—which has worried you so much, Whyte. As for the missing train and the people on it . . ." He gestured them into infinity. "Do you agree, Dr. Tupelo?" he asked the mathematician.

Tupelo shook his head slowly. "Not entirely, Mr. Wilson," he responded. "Now, please keep in mind that I don't fully comprehend what has happened. It's unfortunate that you won't find anybody who can give a good explanation. The one man who might have done so is Professor Turnbull, of Tech, and he was on the train. But in any case, you will want to check my conclusions against those of some competent topologists. I can put you in touch with several.

"Now, with regard to the recovery of the missing train, I can say that I think this is not hopeless. There is a finite probability, as I see it, that the train will eventually pass from the nonspatial part of the network, which it now occupies, back to the spatial part. Since the nonspatial part is wholly inaccessible, there is unfortunately nothing we can do to bring about this transition, or even to predict when or how it will occur. But the possibility of the transition will vanish if the Boylston shuttle is taken out. It is just this section of the track that gives the network its essential singularities. If the singularities are removed, the train can never reappear. Is this clear?"

It was not clear, of course, but the seven listening men nodded agreement. Tupelo continued.

"As for the continued operation of the System while the missing train is in the nonspatial part of the network, I can only give you the facts as I see them and leave to your judgment the difficult decision to be drawn from them. The transition back to the spatial part is unpredictable, as I have already told you. There is no way to know when it will occur, or where. In particular, there is a fifty per cent probability that, if and when the train reappears, it will be running on the wrong track. Then there will be a collision, of course."

The engineer asked: "To rule out this possibility, Dr. Tupelo, couldn't we leave the Boylston shuttle open, but send no trains through it? Then, when the missing train reappears on the shuttle, it cannot meet another train."

"That precaution would be ineffective, Mr. Kennedy," Tupelo answered. "You see, the train can reappear anywhere on the System. It is true that the System owes its topological complexity to the new shuttle. But, with the shuttle in the System, it is now the whole System that possesses infinite connectivity. In other words, the relevant topological property is a property *derived* from the shuttle, but *belonging to* the whole System. Remember that the train made its first transition at a point between Park and Kendall, more than three miles away from the shuttle.

"There is one question more you will want answered. If you decide to go on operating the System, with the Boylston shuttle left in until the train reappears, can this happen again, to another train? I am not certain of the answer, but I think it is No. I believe an exclusion principle operates here, such that only one train at a time can occupy the nonspatial network."

The physician rose from his seat. "Dr. Tupelo," he began timorously, "when the train does reappear, will the passengers—?"

"I don't know about the people on the train," Tupelo cut in. "The topological theory does not consider such matters." He looked quickly at each of the seven tired, querulous faces before him. "I am sorry, gentlemen," he added, somewhat more gently. "I simply do not know." To Whyte, he added, "I think I can be of no more help tonight. You know where to reach me." And, turning on his heel, he left the car and climbed the stairs. He found dawn spilling over the street, dissolving the shadows of night.

That impromptu conference in a lonely subway car was never reported in the papers. Nor were the full results of the night-long vigil over the dark and twisted tunnels. During the week that followed, Tupelo participated in four more formal conferences with Kelvin

Whyte and certain city officials. At two of these, other topologists were present. Ornstein was imported to Boston from Philadelphia, Kashta from Chicago, and Michaelis from Los Angeles. The mathematicians were unable to reach a consensus. None of the three would fully endorse Tupelo's conclusions, although Kashta indicated that there *might* be something to them. Ornstein averred that a finite network could not possess infinite connectivity, although he could not prove this proposition and could not actually calculate the connectivity of the System. Michaelis expressed his opinion that the affair was a hoax and had nothing whatever to do with the topology of the System. He insisted that if the train could not be found on the System then the System must be open, or at least must once have been open.

But the more deeply Tupelo analyzed the problem, the more fully he was convinced of the essential correctness of his first analysis. From the point of view of topology, the System soon suggested whole families of multiple-valued networks, each with an infinite number of infinite discontinuities. But a definitive discussion of these new spatio-hyperspatial networks somehow eluded him. He gave the subject his full attention for only a week. Then his other duties compelled him to lay the analysis aside. He resolved to go back to the problem later in the spring, after courses were over.

Meanwhile, the System was operated as though nothing untoward had happened. The general manager and the mayor's representative had somehow managed to forget the night of the search, or at least to reinterpret what they had seen and not seen. The newspapers and the public at large speculated wildly, and they kept continuing pressure on Whyte. A number of suits were filed against the System on behalf of persons who had lost a relative. The State stepped into the affair and prepared its own thorough investigation. Recriminations were sounded in the halls of Congress. A garbled version of Tupelo's theory eventually found its way into the press. He ignored it, and it was soon forgotten.

The weeks passed, and then a month. The State's investigation was completed. The newspaper stories moved from the first page to the second; to the twenty-third; and then stopped. The missing persons did not return. In the large, they were no longer missed.

One day in mid-April, Tupelo traveled by subway again, from Charles Street to Harvard. He sat stiffly in the front of the first car, and watched the tracks and gray tunnel walls hurl themselves at the train. Twice the train stopped for a red light, and Tupelo found himself wondering whether the other train was really just ahead, or just beyond space. He half hoped, out of curiosity, that his exclusion principle was wrong, that the train might make the transition. But he arrived at Harvard on time. Only he among the passengers had found the trip exciting.

The next week he made another trip by subway, and again the next. As experiments, they were unsuccessful, and much less tense than the first ride in mid-April. Tupelo began to doubt his own analysis. Sometime in May, he reverted to the practice of commuting by subway between his Beacon Hill apartment and his office at Harvard. His mind stopped racing down the knotted gray caverns ahead of the train. He read the morning newspaper, or the abstracts in *Reviews of Modern Mathematics*.

Then there was one morning when he looked up from the newspaper and sensed something. He pushed panic back on its stiff, quivering spring, and looked quickly out the window at his right. The lights of the car showed the black and gray lines of wall-spots streaking by. The tracks ground out their familiar steely dissonance. The train rounded a curve and crossed a junction that he remembered. Swiftly, he recalled boarding the train at Charles, noting the girl on the ice-carnival poster at Kendall, meeting the southbound train going into Central.

He looked at the man sitting beside him, with a lunch pail on his lap. The other seats were filled, and there were a dozen or so straphangers. A mealy-faced youth near the front door smoked a cigarette, in violation of the rules. Two girls behind him across the aisle were discussing a club meeting. In the seat ahead, a young woman was scolding her little son. The man on the aisle, in the seat ahead of that, was reading the paper. The Transit Ad above him extolled Florida oranges.

He looked again at the man two seats ahead and fought down the terror within. He studied that man. What was it? Brunet, graying hair; a roundish head; wan complexion; rather flat features; a thick neck, with the hairline a little low, a little ragged; a gray, pin-stripe suit. While Tupelo watched, the man waved a fly away from his left ear. He swayed a little with the train. His newspaper was folded vertically down the middle. His *newspaper!* It was last March's!

Tupelo's eyes swiveled to the man beside him. Below his lunch pail was a paper. Today's. He turned in his seat and looked behind him. A young man held the *Transcript* open to the sport pages. The date was March 4. Tupelo's eyes raced up and down the aisle. There were a dozen passengers carrying papers ten weeks old.

Tupelo lunged out of his seat. The man on the aisle muttered a curse as the mathematician crowded in front of him. He crossed the aisle in a bound and pulled the cord above the windows. The brakes sawed and screeched at the tracks, and the train ground to a stop. The startled passengers eyed Tupelo with hostility. At the rear of the car, the door flew open and a tall, thin man in a blue uniform burst in. Tupelo spoke first.

"Mr. Dorkin?" he called, vehemently.

The conductor stopped short and groped for words.

"There's been a serious accident, Dorkin," Tupelo said, loudly, to carry over the rising swell of protest from the passengers. "Get Gallagher back here right away!"

Dorkin reached up and pulled the cord four times. "What happened?" he asked.

Tupelo ignored the question, and asked one of his own. "Where have you been, Dorkin?"

The conductor's face was blank. "In the next car, but—"

Tupelo cut him off. He glanced at his watch, then shouted at the passengers. "It's ten minutes to nine on May 17!"

The announcement stilled the rising clamor for a moment. The passengers exchanged bewildered glances.

"Look at your newspapers!" Tupelo shouted. "Your newspapers!"

The passengers began to buzz. As they discovered each other's papers, the voices rose. Tupelo took Dorkin's arm and led him to the rear of the car. "What time is it?" he asked.

"8:21," Dorkin said, looking at his watch.

"Open the door," said Tupelo, motioning ahead. "Let me out. Where's the phone?"

Dorkin followed Tupelo's directions. He pointed to a niche in the tunnel wall a hundred yards ahead. Tupelo vaulted to the ground and raced down the narrow lane between the cars and the wall. "Central Traffic!" he barked at the operator. He waited a few seconds, and saw that a train had stopped at the red signal behind his train. Flashlights were advancing down the tunnel. He saw Gallagher's legs running down the tunnel on the other side of 86. "Get me Whyte!" he commanded, when Central Traffic answered. "Emergency!"

There was a delay. He heard voices rising from the train beside him. The sound was mixed—anger, fear, hysteria.

"Hello!" he shouted. "Hello! Emergency! Get me Whyte!"

"I'll take it," a man's voice said at the other end of the line. "Whyte's busy!"

"Number 86 is back," Tupelo called. "Between Central and Harvard now. Don't know when it made the jump. I caught it at Charles ten minutes ago, and didn't notice it till a minute ago."

The man at the other end gulped hard enough to carry over the telephone. "The passengers?" he croaked.

"All right, the ones that are left," Tupelo said. "Some must have got off already at Kendall and Central."

"Where have they been?"

Tupelo dropped the receiver from his ear and stared at it, his mouth wide open. Then he slammed the receiver onto the hook and ran back to the open door.

Eventually, order was restored, and within a half hour the train proceeded to Harvard. At the station, the police took all passengers into protective custody. Whyte himself arrived at Harvard before the train did. Tupelo found him on the platform.

Whyte motioned weakly toward the passengers. "They're really all right?" he asked.

"Perfectly," said Tupelo. "Don't know they've been gone."

"Any sign of Professor Turnbull?" asked the general manager.

"I didn't see him. He probably got off at Kendall, as usual."

"Too bad," said Whyte. "I'd like to see him!"

"So would I!" Tupelo answered. "By the way, now is the time to close the Boylston shuttle."

"Now is too late," Whyte said. "Train 143 vanished twenty-five minutes ago, between Egleston and Dorchester."

Tupelo stared past Whyte, and down and down the tracks.

"We've got to find Turnbull," Whyte said.

Tupelo looked at Whyte and smiled thinly.

"Do you really think Turnbull got off this train at Kendall?" he asked.

"Of course!" answered Whyte. "Where else?"

KURD LASSWITZ

# The Universal Library

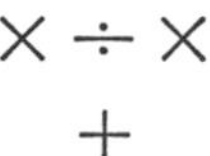

"Come and sit down over here, Max," said Professor Wallhausen, "and stop digging around on my desk. I assure you there is nothing there which you could use for your magazine."

Max Burkel walked over to the living-room table, sat down slowly and reached for his beer glass. "Well, *prosit*, old boy. Nice to be here again. But no matter what you say, you've got to write something for me."

"Unfortunately I don't have any good ideas right now. Besides, so much superfluous stuff is being written and, unfortunately, printed too—"

"You don't have to tell that to a harassed editor like yours truly. The question is, however, just what is superfluous stuff? The authors and their public completely fail to agree about that. Same for editors and reviewers. Well, my three weeks' vacation is just beginning. In the meantime my assistant can do the worrying."

"I have sometimes wondered," said Mrs. Wallhausen, "that you can still find something new for the printer. I should think that by now practically everything that can be expressed with letters has been tried."

"One would think so, but the human mind seems to be inexhaustible—"

"In repetitions, you mean."

"Well, yes," Burkel admitted, "but also when it comes to new ideas and expressions."

"Just the same," mused Professor Wallhausen, "one could express

[REPRINTED BY PERMISSION OF WILLY LEY]

*Authorized translation by Willy Ley*

in print everything which can ever be given to humanity, be it historical information, scientific understanding of the laws of nature, poetic imagination and power of expression, or even the teachings of wisdom. Provided, of course, that it can be expressed in words. After all, our books conserve and disseminate the results of thought. But the number of the possible combinations of a given number of letters is limited. Therefore all possible literature must be printable in a finite number of volumes."

"Dear friend," said Burkel, "now you are talking as a mathematician rather than as a philosopher. How can all possible literature, including that of the future, make a finite number of books?"

"I'll figure out in a moment how many volumes would be required to make a Universal Library. Will you—" he turned to his daughter—"hand me a sheet of paper and a pencil from my desk?"

"Bring the logarithm table too," Burkel added dryly.

"Not necessary, not necessary at all," the professor declared. "But now our literary friend has the first word. I ask: If we are frugal and do away with various fonts of type, writing only for a hypothetical reader who is willing to put up with some typographical inconveniences and who is only interested in the meaning—"

"There is no such reader," Burkel said firmly.

"I said 'hypothetical reader.' How many different characters would one need for printing general literature?"

"Well," said Burkel, "let's just stick to the upper- and lower-case letters of the Latin alphabet, the customary punctuation marks and the space that keeps the words apart. That wouldn't be too much. But for scientific works, that's another story. Especially you mathematicians have an enormous number of symbols."

"Which could be replaced by an agreement with small indices like $a_1$, $a_2$, $a_3$ and $a^1$, $a^2$ and $a^3$, adding just twice ten characters. One could even use this system to write words from languages which do not use the Latin alphabet."

"All right. Maybe your hypothetical, or better, your ideal reader will put up with that too. Under these conditions we could probably express everything with, say one hundred different characters."

"Well, well. Now how big do you want each volume?"

"I should think that one can exhaust a theme pretty well with five hundred book pages. Let's say that there are forty lines per page and fifty characters per line, we'll have forty times fifty times five hundred characters per volume, which is—you calculate it."

"One million," said the professor. "Therefore, if we take our one hundred characters, repeat them in any order often enough to fill a volume which has room for one million characters, we'll get a piece of literature of some kind. Now if we produce all possible combinations mechanically we'll ultimately get all the works which ever have been written in the past or can be written in the future."

Burkel slapped his friend's shoulder. "You know, I'm going to subscribe right now. This will furnish me with all the future volumes of my magazine; I won't have to read manuscripts any more. This is wonderful for both editor and publisher: the elimination of the author from the literary business! The replacement of the writer by the automatic printing press! A triumph of technology!"

"What?" said Mrs. Wallhausen. "You say everything will be in that library? The complete works of Goethe? The Bible? The works of all the classical philosophers?"

"Yes, and with all the variations of wording nobody has thought up yet. You'll find the lost works of Tacitus and their translations into all living and dead languages. Furthermore, all of my and friend Burkel's future works, all forgotten and still undelivered speeches in all parliaments, the official version of the Universal Declaration of Peace, the history of the subsequent wars, all the compositions all of us wrote in school and college—"

"I wish I had had this volume when I was in college," Mrs. Wallhausen said. "Or would it be volumes?"

"Volumes, probably. Don't forget that the space between words is a typographical character too. A book may contain a single line; everything else might be empty. On the other hand, even the longest works could be accommodated because if they don't fit into one volume they could be continued through several."

"No, thanks. Finding something must be a chore."

"Yes, this is one of the difficulties," Professor Wallhausen said with a pleased smile, looking after the smoke from his cigar. "At first glance one should think that this would be simplified by the fact that the library must contain its own catalogue and index—"

"Good!"

"The problem would be to find that one. Moreover, if you had found an index volume it wouldn't help you any because the contents of the Universal Library are not only indexed correctly, but also in every possible incorrect and misleading manner."

"The devil! But unfortunately true."

"Yes, there would be quite a number of difficulties. Let's say we take the first volume of the Universal Library. Its first page is empty, so is the second and the third and so forth through all five hundred pages. This is the volume in which the 'space' has been repeated one million times."

"At least that volume can't contain any nonsense," Mrs. Wallhausen observed.

"Hardly a consolation. But we'll take the second volume. Also empty until, on page 500, line 40 at the extreme end, there is a lonely little 'a.' Same thing in the third volume, but the 'a' has moved up one place. And then the 'a' slowly moves up, place by place, through the first million volumes until it reaches the first place on page 1, line 1 of the first volume of the second million. Things continue that way through the first hundred million volumes until each of the hundred characters has made its lonely way from last to first place in the books. The same then happens with 'aa' or with any other two characters. One volume could contain one million period marks and another one million question marks."

"Well," said Burkel, "it should be simple to recognize and discard those volumes."

"Maybe, but the worst is yet to come. It happens when you have found a volume which seems to make sense. Say you want to refresh your memory about a passage in Goethe's *Faust* and you manage to locate a volume with the right beginning. But when you have progressed for a page or two it goes on 'aaaaaa' and that is the only thing in the remaining pages of the book. Or you find a table of logarithms, but you can't tell whether it is correct. Remember, the Universal Library contains everything which is correct, but also everything which is not. You can't trust the chapter headings either. A volume may begin with the words 'History of the Thirty Years War' and then say: "After the nuptials of Prince Blücher and the Queen of Dahomey had been celebrated at Thermopylae"—you see what I mean. Of course, nobody would ever be embarrassed. If an author has written the most incredible nonsense it will, of course, be in the Universal Library. It will be under his by-line. But it will also be under the by-line of William Shakespeare and under any other possible by-line. He will find one of his books where it is asserted after every sentence that all this is nonsense and another one where it is stated after the selfsame sentences that they are the purest wisdom."

"I have enough," said Burkel. "I knew as soon as you started that this was gong to be a tall tale. I won't subscribe to your Universal Library. It would be impossible to sift truth from falsehood, sense from nonsense. If I find several million volumes all claiming to be the true history of Germany during the twentieth century and all contradicting each other, I would do much better by reading the original works of the historians."

"Very clever! Otherwise you would have taken on an impossible burden. But I wasn't telling a tall tale in your sense. I did not claim that you could use the Universal Library, I merely said that it is possible to tell exactly how many volumes would be required for a Universal Library containing all possible literature."

"Go ahead and calculate," said Mrs. Wallhausen; "it is easy to see that this blank sheet of paper bothers you."

"Not needed," said the professor; "*that* I can do in my head. All we have to do is to realize very clearly how that library is going to be produced. First we put down each one of our hundred characters. Then we add to each, every one of our hundred characters, so that we have one hundred times one hundred groups of two characters each. Adding the third set of our hundred characters we get $100 \times 100 \times 100$ groups of three characters each and so forth. Since we have one million possible positions per volume, the total number of volumes is 100 raised to the millionth power. Now since 100 is the square of 10, we obtain the same figure if we write a '10' with two million as the power. This is simply a '1' followed by two million zeros. Here it is: $10^{2,000,000}$."

"You make your life easy," remarked Mrs. Wallhausen. "Why don't you write it down in the normal manner?"

"Not me. This would take me at least two weeks, without time out for food and sleep. If you printed that figure, it would be a little over two miles long."

"What is the name of that figure?" the daughter wanted to know.

"It has no name. There isn't even a way in which we could hope to grasp that figure, it is so colossal, even though it is finite."

"How about expressing it in trillions?" asked Burkel.

"A mathematical trillion is a nice big figure, a '1' followed by 18 zeros. But if you express the number of volumes in trillions, you get a figure with 1,999,982 zeros instead of two million zeros. It's no help; one is as ungraspable as the other. But, just wait a moment." The professor scribbled a few figures on the sheet of paper.

"I *knew* it would come to that!" Mrs. Wallhausen said with much satisfaction.

"All done," her husband announced. "I assumed that each volume is two centimeters thick and that the whole library is arranged in one long row. How long do you think this row would be?"

"I know," said the daughter. "You want me to say it?"

"Go ahead."

"Twice as many centimeters as the number of volumes."

"Bravo, my dear. Absolutely correct. Now let's look at this more closely. You know that the speed of light, expressed in metric units, is 300,000 kilometers per second, which for a year amounts to about 10,000 million kilometers, which equals 1,000,000,000,000,000,000 centimeters, your mathematical trillion, Burkel. If our librarian can move with the speed of light it will still take him two years to pass a trillion volumes. To go from one end of the library to the other with the speed of light will take twice as many years as there are trillions of volumes in the library. We had that figure before, and I feel that nothing shows more clearly how impossible it is to grasp the meaning of this $10^{2,000,000}$ even though, as I have said repeatedly, it is a finite figure."

"If the ladies will permit, I have one more question," said Burkel. "I suspect that you have calculated a library for which there is no room in the universe."

"We'll see in a moment," the professor answered, reaching for the pencil. "Well, I assumed you packed the library in 1000-volume boxes, each box having a capacity of precisely one cubic meter. All space to the farthest known spiral galaxies would not hold the Universal Library. In fact, you would need this volume of space so often that the number of packed universes would be a figure with only some 60 zeros less than the figure for the number of volumes. No matter how we try to visualize it, we are bound to fail."

"I thought all along that it was infinite," said Burkel.

"No, that's just the point. The figure is not infinite, it is a finite figure. The mathematics of it are flawless. What is surprising is that we can write down on a very small piece of paper the number of volumes comprising all possible literature, something which at first glance seems to be infinite. But if we then try to visualize it—for example, try to find a specific volume—we realize that we cannot grasp what is otherwise a very clear and logical thought which we evolved ourselves."

"Well," concluded Burkel, "coincidence plays, but reason creates. And for this reason you'll write down tomorrow what amused us tonight. That way I'll get an article for my magazine which I can carry with me."

"All right. I'll write it down for you. But I'm telling you right now that your readers will conclude that this is an excerpt from one of the superfluous volumes of the Universal Library."

WILLY LEY

# Postscript to "The Universal Library"

× ÷ ×
+

WHEN LASSWITZ wrote his story about the Universal Library (it was published in a book of short stories in 1901) he did not evolve the whole of the idea. He mainly provided the mathematics of the case which fell into his field, since he was a professor of mathematics in addition to being a philosopher. The earliest version of the idea is connected with the name of Ramon Lully, latinized as Raimundus Lullus, a Spaniard from the island of Majorca who lived from 1235-1315. Lullus was both a missionary and a mystic philosopher and at one time he had an idea which later was referred to as Lully's Great Art, or, also latinized, as *ars magna Lulli*.

The idea, or better, one aspect of it, was about as follows: If I pick out one characteristic of something and state all the possibilities I must, of necessity, state the truth too. For example, the list: blood is blue, blood is green, blood is purple, blood is colorless, blood is black, etc., etc., must contain the correct statement, but this one list alone does not point out which statement is the truth. However, it might be possible to construct other lists of possibilities which will eliminate some of the color possibilities. Therefore, if the whole thing is handled correctly, the truth, that blood is red, should be the only color possibility left over. Hence one would have arrived at a correct statement by means of several lists of statements *which might be constructed mechanically*.

Lullus even tried to build such a device. He used a number of concentric rings which could be turned, thereby bringing words inscribed

[REPRINTED BY PERMISSION OF THE AUTHOR]

on them into new arrangements. A century of experimentation with this and similar devices brought the conclusion—quite obvious to us after a few moments of reflection—that the machine did not succeed in obviating the need for thought in the experimenter. Most of the time it would be impossible to arrive even at a single answer. To use our example again, the machine might leave the three choices: "blood is red," "blood is yellow" and "blood is white," and the experimenter would have to know (or to find out) which choice is correct. But if the machine should end up with the same three choices about the color of roses the observer would have to make a different decision: namely that all three are true.

In retrospect it can now be said that the *ars magna Lulli* was the first seed of what is now called "symbolic logic," but it took a long time until the seed brought fruit, this particular fruit. Other results came about earlier. Among the famous people who spent much thought on the *ars magna Lulli,* which in the end was to make thought superfluous, were Giordano Bruno, Athanasius Kircher and, somewhat later, Gottfried Wilhelm Leibnitz. Leibnitz, influenced by other ideas he was studying at the same time, seems to have reasoned that the machine should be possible and workable if the variety of questions was small, the number of possibilities not too large and, most important of all, each possibility was something definite, like "6" or "12" or "17" and not something vague like the shades of a color. Leibnitz built a calculating machine for adding, subtracting, multiplying, dividing and extracting roots which he exhibited in London in 1763. Whether Pascal, who had built a simpler calculating machine at an earlier date, had also been influenced by Lullus is not known to me.

About a century after the exhibition of Leibnitz' machine, the idea came up that while Lully's original assumption could not be made to work with statements and concepts, it might apply to letters which, in combination, express statements and concepts. The first to actually say so seems to have been the psychologist (and philosopher) Gustav Theodor Fechner, who died in 1887. Lasswitz, at any event, had the idea from Fechner. Of course Lasswitz killed it off beautifully, and I think that the whole case is remarkable and possibly even unique: a philosophical idea wiped out by mathematical treatment presented in the form of fiction!

But then, after this had taken place, the mathematicians became interested. Of course there was no quibbling about the mathematics of the Universal Library, but maybe one could change the assumptions around a bit. At any event it would be interesting to see what would happen.

One simplification was suggested by Dr. Th. Wolff in 1929. Instead of presuming volumes of a million positions one might assume sheets of only 1000 positions each. That would reduce the over-all bulk considerably, especially as those volumes given over to just one letter would be merely single sheets. Sensible things which did not fit a sheet could still go on through any number of sheets. Doctor Wolff also reduced the number of characters. There is no need for capitals. Nor is there any for figures because one can express "4" just as well by writing "four." Likewise, one could dispense with some letters: it would do little harm if both "i" and "j" were expressed by "i," while the "w" could be expressed either by "uu" which corresponds with the name of the discarded letter, or by "vv" which corresponds with its shape. One could do without the "q" and, if necessary, express the "z" by "sc" or by "cs." Similarly the ";" could be expressed by ".," and the ":" by ".." and signs like "+" and "−" could be spelled out as plus and minus.

With such drastic cutting one could make 25 characters do the trick. The final result of 25 characters on 1000-position sheets would be $25^{1000}$ (don't try to write it "in full," you'd need 1398 figures). This is, of course, considerably less than the Lasswitz Universal Library as a simple comparison of the two figures

$10^{2,000,000}$ books

and

$25^{1,000}$ sheets

shows. Still, if you tried to visualize the number of sheets in about the manner described by Lasswitz, you would find that $25^{1000}$ sheets are about as bad as $10^{2,000,000}$ books.

A still later stab at the same idea is the one made by Prof. George Gamow in his book *One, Two, Three . . . Infinity*. In some respects Gamow went farther than either Lasswitz or Wolff, for he even gave some advice as to the engineering design of the automatic printing press.

He said that we have 26 letters in the English alphabet plus 10 figures and 14 common signs and punctuation marks. Now we take a wheel and place these 50 characters, one of which is a blank space, around its rim. We manufacture 65 such wheels, since we are going to use an endless roll of paper like a newsprint roll which has 65 positions to the line. The wheels would be geared in such a way that after the first impression from "standing position" has been made (resulting in a line of either 65 blank spaces or 65 times the letter "a") the first wheel would move forward one position. When the first wheel has made a complete revolution, the next wheel is taken along for one character.

After each turning movement of any one of the wheels, one line is automatically printed. In describing this Gamow showed that such a press could actually be built; in fact that it would not even need special ingenuity. But don't expect any results. The total number of possible different lines is, with 50 characters and 65 positions, $50^{65}$ which is the same as $10^{110}$. Compared to Lasswitz and to Wolff this looks positively small.

That it is not has been shown by Gamow himself:

> To feel the immensity of that number, assume that each atom in the universe represents a separate printing press, so that we have 3 times $10^{74}$ machines working simultaneously. Assume further that all these machines have been working continuously since the creation of the universe, that is for the period of 3000 million years or $10^{17}$ seconds, printing at the rate of atomic vibrations, that is $10^{15}$ lines a second. By now they would have printed about
>
> 3 times $10^{74}$ times $10^{17}$ times $10^{15}$ = 3 times $10^{106}$ lines; which is only about one thirtieth of one per cent of the total number required.

Well, that's that. The "elimination of the author from the literary business" has been proved to be a glorious failure. The editor will still have to read manuscripts. And if he feels, on occasion, that he may have come across some of the "superfluous volumes" from the Universal Library, he might console himself with the thought that the "real thing" would be, in the absolute meaning of that word, immeasurably worse.

HARRY STEPHEN KEELER

—

# John Jones's Dollar

× ÷ ×
+

ON THE 201ST DAY of the year 3221 A.D., the professor of history at the University of Terra seated himself in front of the Visaphone and prepared to deliver the daily lecture to his class, the members of which resided in different portions of the earth.

The instrument before which he seated himself was very like a great window sash, on account of the fact that there were three or four hundred frosted glass squares visible. In a space at the center, not occupied by any of these glass squares, was a dark oblong area and a ledge holding a piece of chalk. And above the area was a huge brass cylinder. Toward this brass cylinder the professor would soon direct his subsequent remarks.

In order to assure himself that it was time to press the button which would notify the members of the class in history to approach their local Visaphones, the professor withdrew from his vest pocket a small contrivance which he held to his ear. As he moved a tiny switch attached to the instrument, a metallic voice, seeming to come from somewhere in space, repeated mechanically: "Fifteen o'clock and one minute—fifteen o'clock and one minute—fifteen o'clock and one min—" Quickly, the professor replaced the instrument in his vest pocket and pressed a button at the side of the Visaphone.

As though in answer to the summons, the frosted squares began, one by one, to show the faces and shoulders of a peculiar type of young men; young men with great bulging foreheads, bald, toothless, and wearing immense horn spectacles. One square, however, still remained empty. On noticing this, a look of irritation passed over the professor's countenance.

But, seeing that every other glass square but this one was filled up, he commenced to talk.

"I am pleased, gentlemen, to see you all posted at your local Visaphones this afternoon. I have prepared my lecture today upon a subject which is, perhaps, of more economic interest than historical. Unlike the previous lectures, my talk will not confine itself to the happenings of a few years, but will gradually embrace the course of ten centuries, the ten centuries, in fact, which terminated three hundred years before the present date. My lecture will be an exposition of the effects of the John Jones Dollar, originally deposited in the dawn of civilization, or to be more precise, in the year of 1921—just thirteen hundred years ago. This John Jon—"

At this point in the professor's lecture, the frosted glass square which hitherto had shown no image, now filled up. Sternly he gazed at the head and shoulders that had just appeared.

"B262H72476Male, you are late to class again. What excuse have you to offer today?"

From the hollow cylinder emanated a shrill voice, while the lips of the picture on the glass square moved in unison with the words:

"Professor, you will perceive by consulting your class book, that I have recently taken up my residence near the North Pole. For some reason, wireless communication between the Central Energy Station and all points north of 89 degrees was cut off a while ago, on account of which fact I could not appear in the Visaphone. Hence—"

"Enough, sir," roared the professor. "Always ready with an excuse, B262H72476Male. I shall immediately investigate your tale."

From his coat pocket the professor withdrew an instrument which, although supplied with an earpiece and a mouthpiece, had no wires whatever attached. Raising it to his lips, he spoke:

"Hello. Central Energy Station, please." A pause ensued. "Central Energy Station? This is the professor of history at the University of Terra speaking. One of my students informs me that the North Pole region was out of communication with the Visaphone System this morning. Is that statement true? I would—"

A voice, apparently from nowhere, spoke into the professor's ear. "Quite true, Professor. A train of our ether waves accidently fell into parallelism with a train of waves from the Venus Substation. By the most peculiar mischance, the two trains happened to be displaced, with reference to each other, one half of a wave length, with the unfortunate result that the negative points of one coincided with the positive points of maximum amplitude of the other. Hence the two wave trains nullified each other and communication ceased for one hundred and eighty-five seconds—until the earth had revolved far enough to throw them out of parallelism."

"Ah! Thank you," replied the professor. He dropped his instrument into his coat pocket and gazed in the direction of the glass square whose image had so aroused his ire. "I apologize, B262H72476Male, for my suspicions as to your veracity—but I had in mind several former experiences." He shook a warning forefinger. "I will now resume my talk.

"A moment ago, gentlemen, I mentioned the John Jones Dollar. Some of you who have just enrolled with the class will undoubtedly say to yourselves: 'What is a John Jones? What is a Dollar?'

"In the early days, before the present scientific registration of human beings was instituted by the National Eugenics Society, man went around under a crude multi-reduplicative system of nomenclature. Under this system there were actually more John Joneses than there are calories in a British Thermal Unit. But there was one John Jones, in particular, living in the twentieth century, to whom I shall refer in my lecture. Not much is known of his personal life except that he was an ardent socialist—a bitter enemy, in fact, of the private ownership of wealth.

"Now as to the Dollar. At this day, when the Psycho-Erg, a combination of the Psych, the unit of esthetic satisfaction, and the Erg, the unit of mechanical energy, is recognized as the true unit of value, it seems difficult to believe that in the twentieth century and for more than ten centuries thereafter, the Dollar, a metallic circular disk, was being passed from hand to hand in exchange for the essentials of life.

"But nevertheless, such was the case. Man exchanged his mental or physical energy for these Dollars. He then re-exchanged the Dollars for sustenance, raiment, pleasure, and operations for the removal of the vermiform appendix.

"A great many individuals, however, deposited their Dollars in a stronghold called a bank. These banks invested the Dollars in loans and commercial enterprises, with the result that, every time the earth traversed the solar ecliptic, the banks compelled each borrower to repay, or to acknowledge as due, the original loan, plus six one-hundredths of that loan. And to the depositor, the banks paid three one-hundredths of the deposited Dollars for the use of the disks. This was known as three per cent, or bank interest.

"Now, the safety of Dollars, when deposited in banks, was not absolutely assured to the depositor. At times the custodians of these Dollars were wont to appropriate them and proceed to portions of the earth, sparsely inhabited and accessible with difficulty. And at other times, nomadic groups known as 'yeggmen' visited the banks, opened the vaults by force, and departed, carrying with them the contents.

"But to return to our subject. In the year 1921, one of these numerous John Joneses performed an apparently inconsequential action which caused the name of John Jones to go down in history. What did he do?

"He proceeded to one of these banks, known at that time as 'The First National Bank of Chicago,' and deposited there one of these disks—a silver Dollar—to the credit of a certain individual. And this individual to whose credit the Dollar was deposited was no other person than the fortieth descendant of John Jones, who stipulated in paper which was placed in the files of the bank that the descendancy was to take place along the oldest child of each of the generations which would constitute his posterity.

"The bank accepted the Dollar under that understanding, together with another condition imposed by this John Jones, namely, that the interest was to be compounded annually. That meant that at the close of each year, the bank was to credit the account of John Jones's fortieth descendant with three one-hundredths of the account as it stood at the beginning of the year.

"History tells us little more concerning this John Jones—only that he died in the year 1931, or ten years afterward, leaving several children.

"Now you gentlemen who are taking mathematics under Professor L127M72421Male, of the University of Mars, will remember that where any number such as x, in passing through a progressive cycle of change, grows at the end of that cycle by a proportion $p$, then the value of the original x, after $n$ cycles, becomes $x\ (1 + p)^n$.

"Obviously, in this case, x equaled one Dollar; $p$ equaled three one-hundredths; and $n$ will depend upon any number of years which we care to consider following the date of deposit. By a simple calculation, those of you who are today mentally alert can check up the results that I shall set forth in my lecture.

"At the time that John Jones died, the amount in the First National Bank of Chicago to the credit of John Jones the fortieth, was as follows."

The professor seized the chalk and wrote rapidly upon the oblong space:

1931 10 years elapsed $1.34

"The peculiar sinuous hieroglyphic," he explained, "is an ideograph representing the Dollar.

"Well, gentlemen, time went on as time will, until a hundred years had passed by. The First National Bank still existed, and the locality, Chicago, had become the largest center of population upon the earth. Through the investments which had taken place, and the yearly compounding of interest, the status of John Jones's deposit was now as follows." He wrote:

2021 100 years elapsed $19.10

"In the following century, many minor changes, of course, took place in man's mode of living; but the so-called socialists still agitated widely for the cessation of private ownership of wealth; the First National Bank still accepted Dollars for safe keeping, and the John Jones Dollar still continued to grow. With about thirty-four generations yet to come, the account now stood:

2121 200 years elapsed $364

"And by the end of the succeeding hundred years, it had grown to what constituted an appreciable bit of exchange value in those days—thus:

2221 300 years $6,920

"Now the century which followed contains an important date. The date I am referring to is the year 2299 A.D., or the year in which every human being born upon the globe was registered under a numerical name at the central bureau of the National Eugenics Society. In our future lessons which will treat with that period in detail, I shall ask you to memorize that date.

"The socialists still agitated, fruitlessly, but the First National Bank of Chicago was now the first International Bank of the Earth. And how great had John Jones's Dollar grown? Let us examine the account, both on that important historical date, and also at the close of the 400th year since it was deposited. Look:

| | | |
|---|---|---|
| 2299 | 378 years | $68,900 |
| 2321 | 400 years | $132,000 |

"But, gentlemen, it had not reached the point where it could be termed an unusually large accumulation of wealth. For larger accumulations existed upon the earth. A descendant of a man once known as John D. Rockefeller possessed an accumulation of great size, but as a matter of fact, it was rapidly dwindling as it passed from generation to generation. So, let us travel ahead another one hundred years. During this time, as we learn from our historical and political archives, the socialists began to die out, since they at last realized the utter futility of combating the balance of power. The account, though, now stood:

2421 500 years $2,520,000

"It is hardly necessary for me to make any comment. Those of you who are most astute, and others of you who flunked my course before and are now taking it the second time, of course know what is coming.

"During the age in which this John Jones lived, there lived also a man, a so-called scientist called Metchnikoff. We know, from a study of our vast collection of Egyptian Papyri and Carnegie Library books, that this Metchnikoff promulgated the theory that old age—or rather senility—was caused by colon-bacillus. This fact was later verified. But while he was correct in the etiology of senility, he was crudely primeval in the therapeutics of it.

"He proposed, gentlemen, to combat and kill this bacillus by utilizing the fermented lacteal fluid from a now extinct animal called the cow, models of which you can see at any time at the Solaris Museum."

A chorus of shrill, piping laughter emanated from the brass cylinder. The professor waited until the merriment had subsided and then continued:

"I beg of you, gentlemen, do not smile. This was merely one of the many similar quaint superstitions existing in that age.

"But a real scientist, Professor K122B62411Male, again attacked the problem in the twenty-fifth century. Since the cow was now extinct, he could not waste his valuable time experimenting with fermented cow lacteal fluid. He discovered that the old *v* rays of Radium—the rays which you physicists will remember are not deflected by a magnetic field—were really composed of two sets of rays, which he termed the g rays and the e rays. These last named rays—only when isolated—completely devitalized all colon-bacilli which lay in their path, without in the least affecting the integrity of any interposed organic cells. The great result, as many of you already know, was that the life of man was extended to nearly two hundred years. That, I state unequivocally, was a great century for the human race.

"But I spoke of another happening—one, perhaps, of more interest than importance. I referred to the bank account of John Jones the fortieth. It, gentlemen, had grown to such a prodigious sum that a special bank and board of directors had to be created in order to care for and reinvest it. By scanning the following notation, you will perceive the truth of my statement:

2521 600 years \$47,900,000

"By the year 2621 A.D., two events of stupendous importance took place. There is scarcely a man in this class who has not heard of how Professor P222D29333Male accidentally stumbled upon the scientific fact that the effect of gravity is reversed upon any body which vibrates perpendicularly to the plane of the ecliptic with a frequency which is an even multiple of the logarithm of 2 of the Napierian base 'e.' At once, special vibrating cars were constructed which carried mankind to all planets. That discovery of Professor P222D29333Male did nothing less than open up seven new territories to our inhabitants; namely: Mercury, Venus, Mars, Jupiter, Saturn, Uranus, and Neptune. In the great land rush that ensued, thousands who were previously poor became rich.

"But, gentlemen, land, which so far had been constituted one of the main sources of wealth, was shortly to become valuable for individual golf links only, as it is today, on account of another scientific discovery.

"This second discovery was, in reality, not a discovery, but the perfection of a chemical process, the principles of which had been known for many centuries. I am alluding to the construction of the vast reducing factories, one upon each planet, to which the bodies of all persons who have died on their respective planets are at once shipped by Aerial Express. Since this process is used today, all of you understand the methods employed; how each body is reduced by heat to its component constituents: hydrogen, oxygen, nitrogen, carbon, calcium, phosphorus, and so forth; how these separated constituents are stored in special reservoirs together with the components from thousands of other corpses; how these elements are then synthetically combined into food tablets for those of us who are yet alive—thus completing an endless chain from the dead to the living. Naturally then, agriculture and stock-raising ceased, since the food problem, with which man had coped from time immemorial, was solved. The two direct results were, first—that land lost the inflated values it had possessed when it was necessary for tillage, and second—that men were at last given enough leisure to enter the fields of science and art.

"And as to the John Jones Dollar, which now embraced countless industries and vast territory on the earth, it stood, in value:

2621 700 years $912,000,000

"In truth, gentlemen, it now constituted the largest private fortune on the terrestrial globe. And in that year, 2621 A.D., there were thirteen generations yet to come before John Jones the fortieth would arrive.

"To continue. In the year 2721 A.D., an important political battle was concluded in the Solar System Senate and House of Representatives. I am referring to the great controversy as to whether the Earth's moon was a sufficient menace to interplanetary navigation to warrant its removal. The outcome of the wrangle was that the question was decided in the affirmative. Consequently—

"But I beg your pardon, young men. I occasionally lose sight of the fact that you are not so well informed upon historical matters as myself. Here I am, talking to you about the moon, totally forgetful that many of you are puzzled as to my meaning. I advise all of you who have not yet attended the Solaris Museum on Jupiter to take a trip there some Sunday afternoon. The Interplanetary Suburban Line runs trains every half hour on that day. You will find there a complete working model of the old satellite of the Earth, which, before it was destroyed, furnished this planet light at night through the crude medium of reflection.

"On account of this decision as to the inadvisability of allowing the moon to remain where it was, engineers commenced its removal in the year 2721. Piece by piece, it was chipped away and brought to the Earth in Interplanetary freight cars. These pieces were then propelled by Zoodolite explosive, in the direction of the Milky Way, with a velocity of 11,217 meters per second. This velocity, of course, gave each departing fragment exactly the amount of kinetic energy it required to enable it to overcome the backward pull of the Earth from here to infinity. I dare say those moon-hunks are going yet.

"At the start of the removal of the moon in 2721 A.D., the accumulated wealth of John Jones the fortieth stood:

2721 800 years $17,400,000,000

"Of course, with such a colossal sum at their command, the directors of the fund had made extensive investments on Mars and Venus.

"By the beginning of the twenty-ninth century, or the year 2807 A.D., the moon had been completely hacked away and sent piecemeal into space, the job having required 86 years. I give, herewith, the result of John Jones's Dollar, both at the date when the moon was completely removed and also at the close of the 900th year after its deposit:

| | | |
|---|---|---|
| 2807 | 886 years | $219,000,000,000 |
| 2821 | 900 years | $332,000,000,000 |

"The meaning of those figures, gentlemen, as stated in simple language, was that the John Jones Dollar now comprised practically all the wealth on Earth, Mars, and Venus—with the exception of one university site on each planet, which was, of course, school property.

"And now I will ask you to advance with me to the year 2906 A.D. In this year the directors of the John Jones fund awoke to the fact that they were in a dreadful predicament. According to the agreement under which John Jones deposited his Dollar away back in the year 1921, interest was to be compounded annually at three per cent. In the year 2900 A.D., the thirty-ninth generation of John Jones was alive, being represented by a gentleman named J664M42721Male, who was thirty years of age and engaged to be married to a young lady named T246M42652Female.

"Doubtless you will ask, what was the predicament in which the directors found themselves? Simply this:

"A careful appraisement of the wealth on Neptune, Uranus, Saturn, Jupiter, Mars, Venus, and Mercury, and likewise Earth, together with an accurate calculation of the remaining heat in the Sun and an appraisement of that heat at a very decent valuation per calorie, demonstrated that the total wealth of the Solar System amounted to 6,309,525,241,362.15.

"But unfortunately, a simple computation showed that if Mr. J664M42721Male married Miss T246M42652Female, and was blessed by a child by the year 2921, which year marked the thousandth year since the deposit of the John Jones Dollar, then in that year there would be due the child, the following amount:

| | | |
|---|---|---|
| 2921 | 1,000 years | $6,310,000,000,000 |

"It simply showed beyond all possibility of argument, that by 2921 A.D., we would be $474,758,637.85 shy—that we would be unable to meet the debt to John Jones the fortieth.

"I tell you, gentlemen, the Board of Directors was frantic. Such wild suggestions were put forth as the sending of an expeditionary force to the nearest star in order to capture some other Solar System and thus obtain more territory to make up the deficit. But that project was impossible on account of the number of years that it would have required.

"Visions of immense lawsuits disturbed the slumber of those unfortunate individuals who formed the John Jones Dollar Directorship. But on the brink of one of the biggest civil actions the courts had ever known, something occurred that altered everything."

The professor again withdrew the tiny instrument from his vest pocket, held it to his ear and adjusted the switch. A metallic voice rasped: "Fifteen o'clock and fifty-two minutes—fifteen o'clock and fifty-two minutes—fift—" He replaced the instrument and went on with his talk.

"I must hasten to the conclusion of my lecture, gentlemen, as I have an engagement with Professor C122B24999Male of the University of Saturn at sixteen o'clock. Now, let me see; I was discussing the big civil action that was hanging over the heads of the John Jones Dollar directors.

"Well, this Mr. J664M42721Male, the thirty-ninth descendant of the original John Jones, had a lover's quarrel with Miss T246M42652-Female, which immediately destroyed the probability of their marriage. Neither gave in to the other. Neither ever married. And when Mr. J664M42721Male died in 2946 A.D., of a broken heart, as it was claimed, he was single and childless.

"As a result, there was no one to turn the Solar System over to. Immediately, the Interplanetary Government stepped in and took possession of it. At that instant, of course, private property ceased. In the twinkling of an eye almost, we reached the true socialistic and democratic condition for which man had futilely hoped throughout the ages.

"That is all today, gentlemen. Class is dismissed."

One by one, the faces faded from the Visaphone.

For a moment, the professor stood ruminating.

"A wonderful man, that old socialist, John Jones the first," he said softly to himself, "a farseeing man, a bright man, considering that he lived in such a dark era as the twentieth century. But how nearly his well-contrived scheme went wrong. Suppose that fortieth descendant had been born?"

# III
# FRACTIONS

ARTHUR T. QUILLER-COUCH

# A New Ballad of Sir Patrick Spens

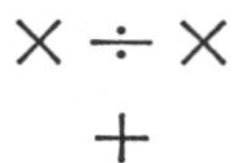

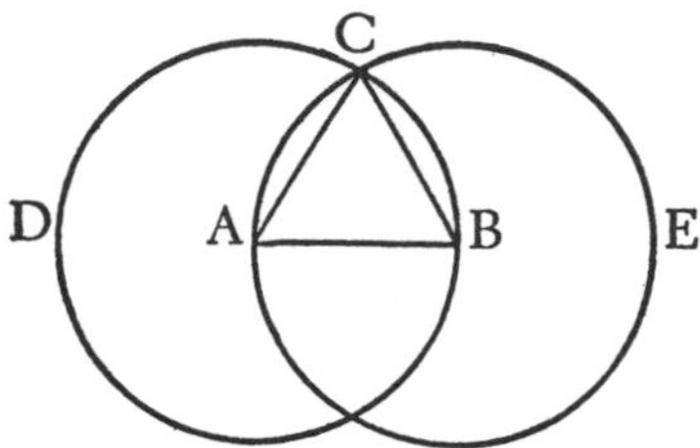


The King sits in Dunfermline town
  Drinking the blude-red wine:
"O wha will rear me an equilateral triangle
  Upon a given straight line?"

O up and spake an eldern knight,
  Sat at the King's right knee—
"Of a' the clerks by Granta side
  Sir Patrick bears the gree.

" 'Tis he was taught by the Tod-huntére
  Tho' not at the tod-huntíng;
Yet gif that he be given a line,
  He'll do as brave a thing."

Our King has written a braid letter
  To Cambrigge or thereby,
And there it found Sir Patrick Spens
  Evaluating $\pi$.

He hadna warked his quotient
  A point but barely three,
There stepped to him a little foot-page
  And louted on his knee.

The first word that Sir Patrick read,
  "*Plus* x" was a' he said:
The neist word that Sir Patrick read,
  'Twas "*plus* expenses paid."

The last word that Sir Patrick read,
  The tear blinded his e'e:
"The pound I most admire is not
  In Scottish currencie."

Stately stepped he east the wa',
  And stately stepped he north:
He fetched a compass frae his ha'
  And stood beside the Forth.

Then gurly grew the waves o' Forth,
  And gurlier by-and-by—
"O never yet was sic a storm,
  Yet it isna sic as I!"

Syne he has crost the Firth o' Forth
  Until Dunfermline toun;
And tho' he came with a kittle wame
  Fu' low he louted doun.

"A line, a line, a gude straight line,
  O King, purvey me quick!
And see it be of thilka kind
  That's neither braid nor thick."

"Nor thick nor braid?" King Jamie said,
  "I'll eat my gude hat-band
If arra line as ye define
  Be found in our Scotland."

## A New Ballad of Sir Patrick Spens

"Tho' there be nane in a' thy rule,
It sall be ruled by me;"
And lichtly with his little pencil
He's ruled the line A B.

Stately stepped he east the wa',
And stately stepped he west;
"Ye touch the button," Sir Patrick said,
"And I sall do the rest."

And he has set his compass foot
Untill the centre A,
From A to B he's stretched it oot—
"Ye Scottish carles, give way!"

Syne he has moved his compass foot
Untill the centre B,
From A to B he's stretched it oot,
And drawn it viz-a-vee.

The tane circle was B C D,
And A C E the tither:
"I rede ye well," Sir Patrick said,
"They interseck ilk ither.

"See here, and whaur they interseck—
To wit, with yon point C—
Ye'll just obsairve that I conneck
The twa points A and B.

"And there ye have a little triangle
As bonny as e'er was seen;
The Whilk is not isosceles,
Nor yet it is scalene."

"The proof! the proof!" King Jamie cried:
"The how and eke they why!"
Sir Patrick laughed within his beard—
" 'Tis *ex hypothesi*—

"When I ligg'd in my mither's wame,
I learn'd it frae my mither,
That things was equal to the same,
Was equal ane to t'ither.

"Sith in the circle first I drew
The lines B A, B C,
By radii true, I wit to you
The baith maun equal be.

"Likewise and in the second circle,
Whilk I drew widdershins,
It is nae skaith the radii baith,
A B, A C, be twins.

"And sith of three a pair agree
That ilk suld equal ane,
By certes they maun equal be
Ilk unto ilk by-lane."

"Now by my faith!" King Jamie saith,
"What *plane* geometrie!
If only Potts had written in Scots,
How loocid Potts wad be!"

"Now wow's my life!" said Jamie the King,
And the Scots lords said the same,
For but it was that envious knicht,
Sir Hughie o' the Graeme.

"Flim-flam, flim-flam!" and "Ho indeed?"
Quod Hughie o' the Graeme;
" 'Tis I could better upon my heid
This prabblin prablem-game."

Sir Patrick Spens was nothing laith
When as he heard "flim-flam,"
By syne he's ta'en a silken claith
And wiped his diagram.

"Gif my small feat may better'd be,
Sir Hew, by they big head,
What I hae done with an A B C
Do thou with X Y Z."

Then sairly sairly swore Sir Hew,
And loudly laucht the King;
But Sir Patrick tuk the pipes and blew,
And *played* that eldritch thing!

## A New Ballad of Sir Patrick Spens

He's play'd it reel, he's played it jig,
And the baith alternative;
And he's danced Sir Hew to the Asses' Brigg,
That's Proposetion Five.

And there they've met, and there they've fet,
Forenenst the Asses' Brigg,
And waefu', waefu', was the fate
That Gar'd them there to ligg.

For there Sir Patrick's slain Sir Hew,
And Sir Hew Sir Patrick Spens—
Now wasna' that a fine to-do
For Euclid's Elemen's?

But let us sing Long live the King!
And his foes the Deil attend 'em:
For he has gotten his little triangle
*Quod erat faciendum!*

CYRIL KORNBLUTH

# The Unfortunate Topologist

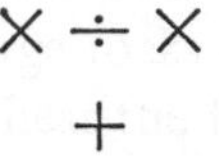

A burleycue dancer, a pip
Named Virginia, could peel in a zip;
  But she read science fiction
  And died of constriction
Attempting a Moebius strip.

[REPRINTED BY PERMISSION OF THE AUTHOR; ORIGINALLY PUBLISHED IN *The Magazine of Fantasy and Science Fiction*]

SIR ARTHUR EDDINGTON

# There Once Was a Breathy Baboon

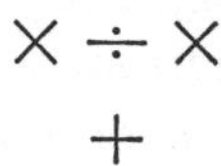

There once was a breathy baboon
Who always breathed down a bassoon,
  For he said, "It appears
  That in billions of years
I shall certainly hit on a tune."

LEWIS CARROLL

—

# Yet What Are All . . .

$\times \div \times$

$+$

Yet what are all such gaieties to me
Whose thoughts are full of indices and surds?

$$x^2 + 7x + 53$$
$$= \frac{11}{3}.$$

RALPH BARTON

# Twinkle, Twinkle, Little Star

Twinkle, twinkle, little star,
How I wonder where you are!
"1.73 seconds of arc from where I seem to be,"
Replied the star, "because $ds^2 = -\left(\frac{1}{2} \cdot \frac{M}{Y}\right)$
$1\, dr^2 - r^2 dO^2 - \left(\frac{1}{2} \cdot \frac{N}{r}\right) d+$."
"Oh," said Arthur, "now I see."

ANDREW MARVELL

—

# Mathematical Love

× ÷ ×
+

As Lines, so Loves *oblique* may well
Themselves in every Angle greet:
But ours so truly Parallel,
Though infinite can never meet.

[FROM *The Definition of Love*]

CHRISTOPHER MORLEY

—

# The Circle

$\times \div \times$
$+$

Few things are perfect: we bear Eden's scar;
Yet faulty man was godlike in design
That day when first, with stick and length of twine,
He drew me on the sand. Then what could mar
His joy in that obedient, mystic line;
Approximating with a zeal divine
He called $\pi$ 3-point-14159
And knew my lovely circuit $2\pi r$!

A circle is a happy thing to be—
Think how the joyful perpendicular
Erected at the kiss of tangency
Must meet my central point, my avatar.
And lovely as I am, yet only 3
Points are needed to determine me.

Q.E.D.

THOMAS DEKKER

# The Circle and the Square

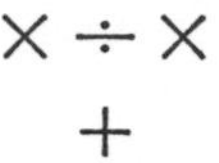

Of geometric figures the most rare,
And perfect'st, are the circle and the square,
The city and the school much build upon
These figures, for both love proportion.
The city-cap is round, the scholar's square,
To show that government and learning are
The perfect'st limbs i' th' body of a state:
For without them, all's disproportionate.

[FROM *The Honest Whore*, PART II, ACT I, SC. 3]

EDNA ST. VINCENT MILLAY

# Euclid Alone Has Looked on Beauty Bare

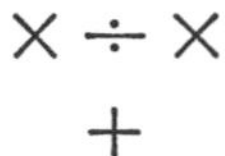

Euclid alone has looked on Beauty bare.
Let all who prate of Beauty hold their peace,
And lay them prone upon the earth and cease
To ponder on themselves the while they stare
At nothing, intricately drawn nowhere
In shapes of shifting lineage; let geese
Gabble and hiss, but heroes seek release
From dusty bondage into luminous air.
O blinding hour, O holy, terrible day,
When first the shaft into his vision shone
Of light anatomized! Euclid alone
Has looked on Beauty bare. Fortunate they
Who, though once only and then but far away,
Have heard her massive sandal set on stone.

VACHEL LINDSAY

# Euclid

× ÷ ×
+

Old Euclid drew a circle
On a sand-beach long ago.
He bounded and enclosed it
With angles thus and so.
His set of solemn graybeards
Nodded and argued much
Of arc and of circumference,
Diameters and such.
A silent child stood by them
From morning until noon
Because they drew such charming
Round pictures of the moon.

A. E. HOUSMAN

# To Think That Two and Two Are Four

× ÷ ×
+

—To think that two and two are four
  And neither five nor three,
The heart of man has long been sore
  And long 'tis like to be.

SAMUEL BUTLER

# The Uses of Mathematics

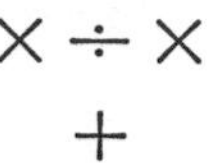

In Mathematicks he was greater
Than Tycho Brahe, or Erra Pater:
For he, by Geometrick scale,
Could take the size of Pots of Ale;
Resolve by Signs and Tangents streight,
If Bread or Butter wanted weight;
And wisely tell what hour o' th' day
The Clock doth strike, by Algebra.

[FROM *Hudibras*, FIRST PART, CANTO I]

CARL SANDBURG

# Arithmetic

Arithmetic is where numbers fly like pigeons in and out of your head.
Arithmetic tells you how many you lose or win if you know how many you had before you lost or won.
Arithmetic is seven eleven all good children go to heaven—or five six bundle of sticks.
Arithmetic is numbers you squeeze from your head to your hand to your pencil to your paper till you get the answer.
Arithmetic is where the answer is right and everything is nice and you can look out of the window and see the blue sky—or the answer is wrong and you have to start all over and try again and see how it comes out this time.
If you take a number and double it and double it again and then double it a few more times, the number gets bigger and bigger and goes higher and higher and only arithmetic can tell you what the number is when you decide to quit doubling.
Arithmetic is where you have to multiply—and you carry the multiplication table in your head and hope you won't lose it.
If you have two animal crackers, one good and one bad, and you eat one and a striped zebra with streaks all over him eats the other, how many animal crackers will you have if somebody offers you five six seven and you say No no no and you say Nay nay nay and you say Nix nix nix?
If you ask your mother for one fried egg for breakfast and she gives you two fried eggs and you eat both of them, who is better in arithmetic, you or your mother?

JOHN ATHERTON

—

# Threes

(TO BE SUNG BY NIELS BOHR)

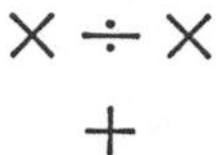

I think that I shall never c
A # lovelier than 3;
For $3<6$ or 4,
And than 1 it's slightly more.

All things in nature come in 3s,
Like ∴s, trios, Q.E.D.s;
While $s gain more dignity
If augmented 3 × 3——

A 3 whose slender curves are pressed
By banks, for compound interest;
Oh, would that, paying loans or rent,
My rates were only 3%!

$3^2$ expands with rapture free,
And reaches toward ∞;
3 complements each x and y,
And intimately lives with π.

A ⊙'s # of °
Are best ÷ up by 3s,
But wrapped in dim obscurity
Is the $\sqrt{-3}$.

Atoms are split by men like me,
But only God is 1 in 3.

EMMA ROUNDS

# Plane Geometry

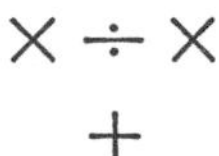

'Twas Euclid, and the theorem pi
  Did plane and solid in the text,
All parallel were the radii,
  And the ang-gulls convex'd.

"Beware the Wentworth-Smith, my son,
  And the Loci that vacillate;
Beware the Axiom, and shun
  The faithless Postulate."

He took his Waterman in hand;
  Long time the proper proof he sought;
Then rested he by the XYZ
  And sat awhile in thought.

And as in inverse thought he sat
  A brilliant proof, in lines of flame,
All neat and trim, it came to him.
  Tangenting as it came.

"AB, CD," reflected he—
  The Waterman went snicker-snack—
He Q.E.D.-ed, and, proud indeed,
  He trapezoided back.

"And hast thou proved the 29th?
  Come to my arms, my radius boy!
O good for you! O one point two!"
  He rhombused in his joy.

'Twas Euclid, and the theorem pi
  Did plane and solid in the text;
All parallel were the radii,
  And the ang-gulls convex'd.

HERBERT DINGLE

—

# He Thought He Saw Electrons Swift

× ÷ ×
+

He thought he saw electrons swift
  Their charge and mass combine.
He looked again and saw it was
  The cosmic sounding line.
"The population then," said he,
  "Must be $10^{79}$."

BRUCE ELLIOTT

—

# Fearsome Fable

$\times \div \times$
$+$

AFTER THEY PUT the fifteen apes in front of the typewriters there was a long wait. The animals sat and looked at the machines, at the paper on the rollers. There was a long pause, then each ape, one after the other, leaned forward and typed a single, different word.

The experimenter waited a long, long time. But after the one flurry of activity, nothing happened. Finally, seeing that the apes had no intention of continuing, he went toward the typewriters.

The first ape had typed NOW; the second had typed IS; the third one, THE; the fourth, TIME; the fifth, FOR; the sixth ape, ALL; the seventh, GOOD; the eighth one, PARTIES; the ninth, TO; the tenth, COME; the eleventh, TO; the twelfth ape, THE; the thirteenth, AID; the fourteenth, OF; and the last ape had typed MAN.

G. H. HARDY

# Bertrand Russell's Dream

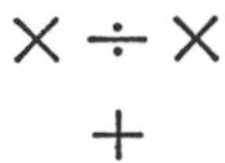

I CAN REMEMBER Bertrand Russell telling me of a horrible dream. He was on the top floor of the University Library, about A.D. 2100. A library assistant was going round the shelves carrying an enormous bucket, taking down book after book, glancing at them, restoring them to the shelves or dumping them into the bucket. At last he came to three large volumes which Russell could recognize as the last surviving copy of *Principia Mathematica*. He took down one of the volumes, turned over a few pages, seemed puzzled for a moment by the curious symbolism, closed the volume, balanced it in his hand and hesitated. . . .

[REPRINTED BY PERMISSION; FROM *A Mathematician's Apology*, © 1948, G. H. HARDY, CAMBRIDGE UNIVERSITY PRESS, LONDON]

C. STANLEY OGILVY

# For All Practical Purposes

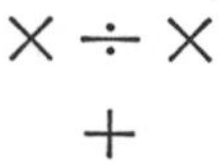

A PROFESSOR, asked what he meant by the phrase ["for all practical purposes"] explained:

"Suppose all the young men in this class were to line up on one side of the room, and all the young ladies on the other. At a given signal, the two lines move toward each other, halving the distance between them. At a second signal, they move forward again, halving the remaining distance; and so on at each succeeding signal. Theoretically, the boys would never reach the girls; but actually, after a relatively small number of moves, they would be close enough for all practical purposes."

[FROM *Through the Mathescope*, © 1956, C. STANLEY OGILVY, OXFORD UNIVERSITY PRESS, LONDON]

## LEWIS CARROLL

# Eternity: A Nightmare

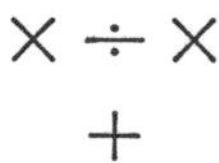

AFTER A MINUTE OR TWO the Earl began again. "If I'm not wearying you, I would like to tell you an idea of the future Life which has haunted me for years, like a sort of waking nightmare—I can't reason myself out of it."

"Pray do," Arthur and I replied, almost in a breath.

"The one idea," the Earl resumed, "that has seemed to me to overshadow all the rest, is that of *Eternity*—involving, as it seems to do, the necessary *exhaustion* of all subjects of human interest. Take Pure Mathematics, for instance—a Science independent of our present surroundings. I have studied it, myself, a little. Take the subject of circles and ellipses—what we call 'curves of the second degree.' In a future Life, it would only be a question of so many years (or *hundreds* of years, if you like) for a man to work out *all* their properties. Then he *might* go to curves of the third degree. Say *that* took ten times as long (you see we have *unlimited* time to deal with). I can hardly imagine his *interest* in the subject holding out even for those; and, though there is no limit to the *degree* of the curves he might study, yet surely the time, needed to exhaust *all* the novelty and interest of the subject, would be absolutely *finite?* And so of all other branches of Science. And, when I transport myself, in thought, through some thousands or millions of years, and fancy myself possessed of as much Science as one created reason can carry, I ask myself, 'What then? With nothing more to learn, can one rest content on *knowledge*, for the eternity yet to be lived through?' It has been a very wearying thought to me. I have sometimes fancied one *might*, in that event, say, 'It is better *not* to be,' and pray for personal *annihilation*—the Nirvana of the Buddhists."

[FROM *Sylvie and Bruno Concluded*, CHAPTER XVI]

GEORGE GAMOW

# An Infinity of Guests

× ÷ ×
+

IN FACT in the world of infinity *a part may be equal to the whole!* This is probably best illustrated by an example taken from one of the stories about the famous German mathematician David Hilbert. They say that in his lectures on infinity he put this paradoxical property of infinite numbers in the following words:

"Let us imagine a hotel with a finite number of rooms, and assume that all the rooms are occupied. A new guest arrives and asks for a room. 'Sorry,' says the proprietor, 'but all the rooms are occupied.' Now let us imagine a hotel with an *infinite* number of rooms, and all the rooms are occupied. To this hotel, too, comes a new guest and asks for a room.

" 'But of course!' exclaims the proprietor, and he moves the person previously occupying room N1 into room N2, the person from room N2 into room N3, the person from room N3 into room N4, and so on. . . . And the new customer receives room N1, which became free as the result of these transpositions.

"Let us imagine now a hotel with an infinite number of rooms, all taken up, and an infinite number of new guests who come in and ask for rooms.

" 'Certainly, gentlemen,' says the proprietor, 'just wait a minute.'

"He moves the occupant of N1 into N2, the occupant of N2 into N4, the occupant of N3 into N6, and so on, and so on. . . .

"Now all odd-numbered rooms become free and the infinity of new guests can easily be accommodated in them."

[REPRINTED BY PERMISSION; FROM *One, Two, Three . . . Infinity*, © 1947, GEORGE GAMOW, VIKING PRESS, INC., NEW YORK]

# SIR ARTHUR EDDINGTON

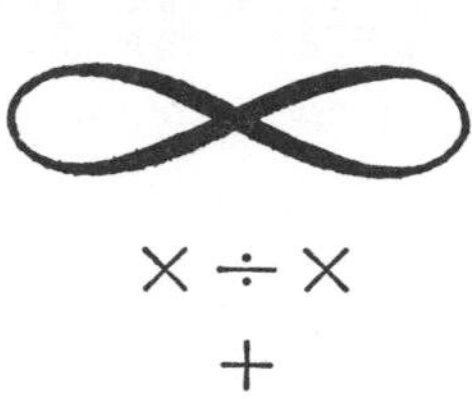

THAT QUEER QUANTITY "infinity" is the very mischief, and no rational physicist should have anything to do with it. Perhaps that is why mathematicians represent it by a sign like a love-knot.

[FROM *New Pathways in Science*]

# No Power on Earth

No power on earth, however great,
Can pull a string, however fine,
Into a horizontal line
That shall be absolutely straight.[1]

[1] This is a slightly improved version of the original, by the learned Master of Trinity College, Dr. William Whewell (1794-1866) who in *An Elementary Treatise on Mechanics* (1819) delivered himself, quite unconsciously, of the following prose sentence: "Hence no force however great can stretch a cord however fine into a horizontal line which is accurately straight."—ED.

# EDGAR ALLAN POE

—

$$(\times + 1)$$

$$\times \div \times$$

$$+$$

[Of two contemporary writers]: "To speak algebraically: Mr. M. [Cornelius Mathews] is execrable but Mr. C. [William Ellery Channing] is $(x + 1)$=ecrable."

[from *James Russell Lowell*]

EDWARD SHANKS

# The Receptive Bosom

$\times \div \times$
$+$

"THE MOST REMARKABLE numerical statement in English literature, Gibbon's 'A thousand swords were plunged at once into the bosom of the unfortunate Probus.' "—in Sunday Times (London), reviewing H. McKay, *The World of Numbers*.

ARTHUR SCHNITZLER

—

# Leinbach's Proof

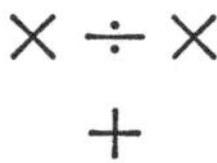

THE STREETS were almost deserted. A steeple clock struck two. It was good, he reflected, that he did not yet have to keep office hours, and that he could sleep late tomorrow. He walked rapidly, surely, humming to himself. Finally he began to sing in a low rich voice that seemed strange to him. Perhaps, indeed, this is not I. Perhaps I am dreaming. Perhaps this is my last dream, the death-bed dream! He remembered an idea that Leinbach once, years ago, had expounded to a large gathering, quite seriously, in fact with a certain impressiveness. Leinbach had discovered a proof that there really is no death. It is beyond question, he had declared, that not only at the moment of drowning, but at all the moments of death of any nature, one lives over again his whole past life with a rapidity inconceivable to others. This remembered life must also have a last moment, and this last moment its own last moment, and so on, and hence, dying is itself Eternity, and hence, in accordance with the theory of limits, one may approach death but can never reach it.

[FROM *Flight into Darkness*]

# Problem

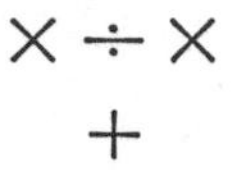

You know those terrible arithmetic problems about how many peaches some people buy, and so forth? Well, here's one we *like*, made up by a third-grader who was asked to think up a problem similar to the ones in his book: "My father is forty-four years old. My dog is eight. If my dog was a human being, he would be fifty-six years old. How old would my father be if he was a dog? How old would my father plus my dog be if they were both human beings?"

# A Letter to Tennyson

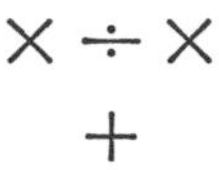

WHEN TENNYSON WROTE "The Vision of Sin," Babbage read it. After doing so, it is said he wrote the following extraordinary letter to the poet:

"In your otherwise beautiful poem, there is a verse which reads:

'Every moment dies a man,
Every moment one is born.'

"It must be manifest that, were this true, the population of the world would be at a standstill. In truth the rate of birth is slightly in excess of that of death. I would suggest that in the next edition of your poem you have it read:

'Every moment dies a man,
Every moment 1 1/16 is born.'

"Strictly speaking this is not correct. The actual figure is a decimal so long that I cannot get it in the line, but I believe 1 1/16 will be sufficiently accurate for poetry. I am, etc."

[FROM THE *Mathematical Gazette*]

# A Fable

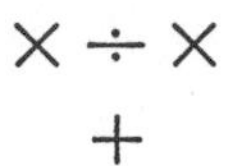

**Once upon a time** there was a teacher who set his class an examination to perform. And when the youths had finished he marked their scripts. But at the end of his labors he found that, by evil chance, he had worked with a total of 99. And, being an industrious man, he converted all the marks into percentages.

So it was that a pupil with 58 marks gained

58.585858 . . . per cent

and a pupil with 73 marks gained

73·737373 . . . per cent,

and others likewise.

And when the time was come that he should return the scripts to his class, being an honest man as well as an industrious, he confessed what he had done and delivered to them their marks in the form of percentages.

Until he came to one named Smith whose work was perfect, to whom perforce he had awarded the percentage

99·999999 . . . per cent.

"So, Smith Minor," saith he, "though I find no fault in you, yet your percentage falls short of the full total of 100. What say you?"

"Sir," saith Smith Minor, moved to anger, "I call that the limit."

[from the *Mathematical Gazette*, vol. 38, 1954, p. 308]

# There Was a Young Man from Trinity

× ÷ ×
+

There was a young man from Trinity,
Who solved the square root of infinity.
  While counting the digits,
  He was seized by the fidgets,
Dropped science, and took up divinity.

—Anonymous

# There Was an Old Man Who Said, "Do"

$\times \div \times$
$+$
$-$

There was an old man who said, "Do
Tell me *how* I should add two and two?
  I think more and more
  That it makes about four—
But I fear that is almost too few."

—Anonymous

$$-$$

# Relativity

$$\times \div \times$$
$$+$$

There was a young lady named Bright,
Who traveled much faster than light.
  She started one day
  In the relative way,
And returned on the previous night.

—ANONYMOUS

—

# There Was a Young Fellow Named Fisk

× ÷ ×
+

There was a young fellow named Fisk
Whose fencing was agile and brisk.
  So fast was his action,
  The Fitz-Gerald contraction
Diminished his sword to a disk.

—Anonymous

# *ACKNOWLEDGMENT OF COPYRIGHT*

*Every reasonable effort has been made to contact the copyright holders for all of the pieces appearing in this anthology.*

Atherton: "Threes (To Be Sung by Niels Bohr)" appeared in *The New Yorker,* March 2, 1957. Reprinted by permission. Copyright © 1957 The New Yorker Magazine, Inc.

Balchin: "God and the Machine" by Nigel Balchin. From *Last Recollections of My Uncle Charles.* Copyright © 1951, 1953, 1954 by Nigel Balchin. Reprinted by permission of Henry Holt & Co., Inc.

Barton: "Twinkle, Twinkle Little Star" by Ralph Barton. From *Science in Rhyme and Without Reason.* Copyright © 1924 by G.P. Putnam's Sons. Reprinted by permission of the publisher.

Clarke: "Superiority" by Arthur C. Clarke. Copyright © 1951 by Fantasy House, Inc. Reprinted by permission of author and author's agents, Scovil Chichak Galen Literary Agency, Inc.

Gamow: "An Infinity of Guests" by George Gamow. From *One, Two, Three . . . Infinity.* Copyright © 1947 by George Gamow. Reprinted by permission of the Estate of George Gamow.

Gardner: "No-Sided Professor" by Martin Gardner. From *The No-Sided Professor and Other Tales* (Amherst, NY: Prometheus Books). Copyright © 1946, renewed © 1987 by Martin Gardner. Reprinted by permission of the publisher.

Hardy: "Bertrand Russell's Dream" by G.H. Hardy. From *A Mathematician's Apology.* Copyright © 1948 by G.H. Hardy. Reprinted by permission of Cambridge University Press.

Housman: "To Think That Two and Two are Four" from "When first my way to fair I took." From *The Collected Poems of A.E. Housman.* Copyright © 1922, 1965 by Henry Holt and Co. Copyright © 1950 by Barclays Bank, Ltd. Reprinted by permission of Henry Holt & Co., Inc.

Huxley: "Young Archimedes: by Aldous Huxley. From *Young Archimedes.* Copyright © 1924, renewed © 1952 by Aldous Huxley. Reprinted by permission of the Estate of Aldous Huxley.

## Acknowledgment of Copyright

Koestler: "Pythogoras and the Psychoanalyst" by Arthur Koestler. From *Arrival and Departure.* Copyright © 19xx by Arthur Koestler. Reprinted by permission of Sterling Lord Literistic.

Lindsay: "Euclid" by Vachel Lindsay. From *Collected Poems of Vachel Lindsay* (New York: Macmillan). Copyright © 1925 Vachel Lindsay. Reprinted by permission of Simon & Schuster, Inc.

Llewellyn: "Mother and the Decimal Point" by Richard Llewellyn. From *How Green Was My Valley.* Copyright © 1940 by Richard Llewellyn. Reprinted by permission of Curtis Brown, Ltd.

Maloney: "Inflexible Logic" appeared in *The New Yorker*, February 3, 1940. Reprinted by permission. Copyright © 1940 by The New Yorker Magazine, Inc.

Millay: "Euclid Alone Has Looked on Beauty Bare" by Edna St. Vincent Millay. From *Collected Poems* (New York: HarperCollins). Copyright © 1923, renewed © 1951 by Edna St. Vincent Millay and Norma Millay Ellis. All rights reserved. Reprinted by permission of Elizabeth Barnett, Literary Executor.

*The New Yorker*: "Problem" from "Talk of the Town" appeared in *The New Yorker*, March 2, 1957. Reprinted by permission. Copyright © 1957 by The New Yorker Magazine, Inc.

Rounds: "Plane Geometry" by Emma Rounds. From *Creative Youth* by Hughes Mearns. Copyright © 1925 by Doubleday, a division of Bantam Doubleday Dell Publishing Group, Inc. Used by permission of Doubleday, a division of Bantam Doubleday Dell Publishing Group, Inc.

Sandburg: "Arithmetic" by Carl Sandburg. From *The Complete Poems of Carl Sandburg.* Copyright © 1950 by Carl Sandburg, renewed © 1978 by Janet Sandburg, Margaret Sandburg, and Helga Sandburg Crile. Reprinted by permission of Harcourt Brace & Company.